Cichlasoma ellioti

Other titles in this series:

The Tropical Aquarium
Community Fishes
Marine Fishes
The Healthy Aquarium
Aquarium Plants
Fish Breeding
African and Asian Catfishes
South American Catfishes
Koi
Livebearing Fishes
Fancy Goldfishes
African Cichlids

White water in southern Costa Rica.

A FISHKEEPER'S GUIDE TO

CENTRAL AMERICAN CICHLIDS

**A detailed survey of this colourful and challenging
group of freshwater tropical fishes**

David Sands

No. 16079

A Salamander Book

ISBN 3–923880–58–8

Cichlasoma labiatum

Credits

Editor: Geoff Rogers Designer: Tony Dominy
Colour reproductions
Melbourne Graphics Ltd.
Filmset: SX Composing Ltd.
Printed in Belgium by Henri Proost & Cie, Turnhout.

Author

A combination of practical experience and academic research has made David Sands ideally qualified to present this survey of Central American cichlids. As an importer and retailer of tropical freshwater and marine fishes, David has encountered at first hand all the challenges which face the fishkeeper. In pursuit of his life-long fascination with fishkeeping, he has made frequent field trips to study and photograph tropical fishes in their natural habitats. A regular contributor to leading aquarist magazines in the UK and the USA, he can advise both accomplished and aspiring fishkeepers how to keep and enjoy these fishes

Consultant

Ian C. Sellick, BSc., is the chairman of the British Cichlid Association. He graduated in Zoology from Bristol University and subsequently spent three years researching cichlid biology. After a period spent with a multinational company as an analyst, he currently works as a consultant for a major German filtration company. He has travelled and collected fish in Central America, and contributes regular articles to several aquarium publications, both on cichlids and a diversity of other aquarium topics. Ian lectures on many subjects, both in the UK and overseas.

Contents

Introduction 8

Aquarium selection 10

Water requirements and filtration 14

Lighting and heating 20

Aquascaping 22

Feeding and routine maintenace 30

Basic health care 34

Breeding 36

Species section 40
A photographic survey of 33 species
of Central American cichlids

Index 76

Picture credits 77

7

Introduction

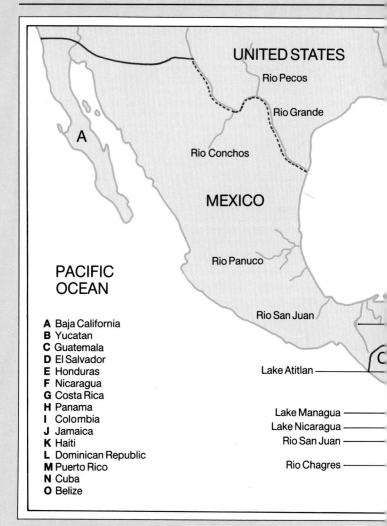

PACIFIC OCEAN

A Baja California
B Yucatan
C Guatemala
D El Salvador
E Honduras
F Nicaragua
G Costa Rica
H Panama
I Colombia
J Jamaica
K Haiti
L Dominican Republic
M Puerto Rico
N Cuba
O Belize

UNITED STATES
Rio Pecos
Rio Grande
Rio Conchos
MEXICO
Rio Panuco
Rio San Juan
Lake Atitlan
Lake Managua
Lake Nicaragua
Rio San Juan
Rio Chagres

To experienced aquarists, Central American cichlids represent an alternative to community fishes; to the impulse keeper of exotic species, they are a challenge. To both groups of fishkeepers, the 'Guapotes' – a name used to describe the giant predators – and their smaller cousins are lively and colourful enough to parallel the popularity of the better known South American or African Rift Lake cichlids. Although the large forms really belong in huge aquariums – and there are those quite willing to maintain display systems – thankfully, many species do not grow quite so big and are easily kept in modest aquariums with medium-sized community fishes.

Most of the species available today have come from North American and German professional collectors who have spawned and raised stock from wild-caught specimens and exported them around the world. In the late 1970s, *Cichlasoma nicaraguense* caught the eye of the aquarist world. In spawning dress, this

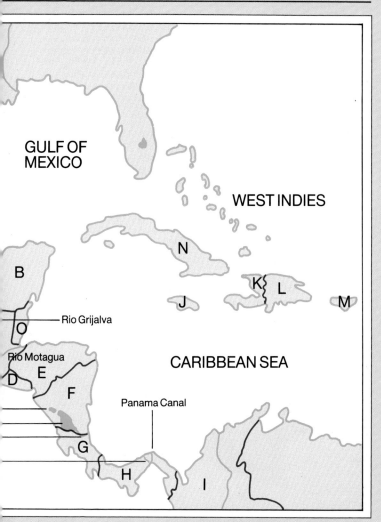

GULF OF
MEXICO

WEST INDIES

N

B

Rio Grijalva

O

Rio Motagua

D   E

F

Panama Canal

CARIBBEAN SEA

J

K L

M

G

H

I

species is ablaze with golden yellow and green, underlined by a black lateral stripe – a splendid display of colour surpassed only by marine reef fishes. Before this and other equally exotic species became available, most enthusiasts knew Central American cichlids through the Red Devil (*Cichlasoma citrinellum*) and the Firemouth Cichlid (*Cichlasoma meeki*) – dramatic names that have helped to create an interest in the so-called 'crater lake fishes'.

Most Central American cichlids can be spawned in aquariums, and a ready market for the fry has helped to stimulate interest still further. General guidance on breeding Central American cichlids is just one of the aspects examined in this practical section of the book and reflected in the detailed species-by-species descriptions beginning on page 42.

The above map of the Central American region highlights the rivers and lakes specifically mentioned in the text of the book.

Aquarium selection

In an ideal world, the selection of the perfect aquarium for Central American cichlids would be determined by the largest space available in the home. For these inhabitants of the rivers, swamps, lagoons and crater lakes that stretch from Texas and Mexico south to Panama need plenty of space in which to exercise various postures, such as territorial defence, sexual display and dominance behaviour. In the waters of the Rio Grande, in Texas and Mexico, or in the Rio Chucunaque, Panama, juvenile cichlids will shoal in great numbers or a spawning pair will defend a pit up to 3m (10ft) across from all-comers that threaten them.

Aquarium compatibility
General fishkeeping practice usually dictates that individual fishes are introduced into the same system over a period of time. The result is that a dominant individual often rules the community, and in certain situations, will relentlessly bully weaker, smaller cichlids. The blame for this brutality is usually placed on the fishes, but this illogical thinking hides the fact that the responsibility really lies with the fishkeeper. It is imperative for the fishkeeper to choose the species to be mixed and the numbers that can be reasonably accommodated within the optimum aquarium size. The first parameter is dictated by the species *available* in local or nearby aquarium shops; the second by the space and funds at hand. If you are considering keeping the territorial minded cichlids, these basic points are crucial.

Central American cichlids divide up reasonably well into small and medium-sized species and the giants, or Guapotes. This division helps to settle the selection decision

Below left: *Despite being the smallest Central American Cichlid,* Neetroplus nematopus *has the distinction of being one of the most aggressive species kept in aquariums.*

Above: Cichlasoma meeki *is an excellent beginner's species to introduce into a small or medium-sized Central American system. It is usually widely available.*

because a small aquarium will accommodate the small to medium-sized species; a large aquarium, the Guapotes.

Small and medium-sized species
The smallest Central American cichlid is *Neetroplus nematopus*, a dwarf species 75-100mm (3-4in) long that is a common, rather secretive inhabitant of the rocky crater lakes of Nicaragua and the Atlantic slope rivers of Costa Rica. Despite its small size, however, it is well able to hold its own among larger cichlids. A dominant male, for example, can destroy the other fishes in a community aquarium and bully cichlids many times its own size. Such aggressive behaviour makes a nonsense of hard and fast rules that directly equate fish size to aquarium size.

Four to six juvenile specimens of *Neetroplus nematopus* could be kept successfully in a suitably aquascaped and filtered aquarium measuring 60×38×30cm (24×15×12in). (Aquarium measurements refer to length, depth and width respectively.) However, it is quite likely that these four or six fishes will eventually

finish up as one or, if luck will have it, a spawning pair. In the larger arena of a crater lake or river, beaten off specimens can retreat to safety, but in the confines of an aquarium they have no escape and will almost certainly be killed by the more aggressive fishes. Housing the same population in an aquarium measuring 90×45×30cm (36×18×12in) would be preferable because it would allow fishes lower down in the 'pecking order' greater opportunities to find sanctuary.

A spawning pair of cichlids will always disrupt a community aquarium. The joy of success in breeding them is often tempered by the damage the pair causes in defending their eggs and fry. Such fierce parental care is the key to survival in their natural habitat and the fishkeeper should only marvel at it and learn to cope with its consequences in the aquarium.

The Firemouth Cichlid, *Cichlasoma meeki* and the Tricolor Cichlid, *C.salvini*, could be kept in similarly sized aquariums, although a slightly larger one measuring 120×45×30cm (48×18×12in) could be of long-term benefit. These medium-sized cichlids (about

11

150mm/6in in length) are widely available from good aquarium dealers. They can be introduced into communities of larger Central American cichlids, providing the aquarium is spacious enough to accommodate the territorial needs of all the fishes.

Giant species
While many rules are made for breaking, the sensible one that states that each giant predatory cichlid should be kept in 137-227 litres (30-50 gallons) of water is not one to be ignored. Correctly fed specimens of *Cichlasoma motaguense* and *C. managuense*, for example, would quickly outgrow and pollute smaller water corridors unless massive regular water changes are carried out and excessive filters are brought into play. At 300mm (12in) in length, both species can be brilliant and colourful in the aquarium, particularly the colour form of *C. managuense* known as the Jaguar Cichlid. Neither species will achieve a bright, keen-eyed and alert condition, however, if they are confined to a small aquarium.

How can these giants be accommodated? Increasing the width (front to back) measurement of the aquarium is one way of gaining much needed space. This also increases the surface area of the water and thus enhances the rate of gaseous exchange in the aquarium. A superb aquarium would be 240×60×60cm (96×24×24in), but a more realistic system would measure 150×45×38cm (60×18×15in). The recommended stocking levels cannot be rigid because a great deal depends on the size and strength of the cichlids; even in the perfect aquarium a large sexually active pair of these cichlids can dominate 30-50 per cent of the system!

The larger *Cichlasoma dovii* and *Petenia splendida* are much rarer in aquarium circles. Perhaps the attainable size of 500-700mm (20-28in) is a deterrent to all but real enthusiasts and public aquariums. The widespread *Cichla ocellaris* is

available from South American rather than Central American stock. This piscine predator grows to a length of 600-700mm (24-28in) and also requires a great deal of swimming space.

Coping with large aquariums
It is clear that the size of the system depends on the choice and the number of specimens to be kept. Some aquarists suggest that the problems of aggression can be solved by overstocking the

Below: *Cichlasoma dovii, a true giant species that attracts a great deal of interest from cichlid enthusiasts. Juveniles can be raised easily in small aquarium systems, but adults need very spacious surroundings.*

aquarium, in much the same way that African Rift Valley Lake Cichlid territorial problems can be overcome. This route is fraught with danger, however, for infections can spread alarmingly as a direct result of overcrowding and it is very difficult to stabilize water chemistry and quality in these situations.

Some experienced breeders crowd juveniles together in the knowledge that the situation is a temporary one until a breeding pair can be separated out. This strategy can also be successful in large community systems, but it demands a great deal of attention and time-consuming maintenance.

The majority of fishkeepers wish to include other species of fishes in a theme aquarium containing cichlids, such as catfishes, characins and, in some cases, livebearers. The assumption that the larger the system the more fishes can be kept and the easier it is to maintain, applies only if the fishkeeper can afford to incorporate the required filtration systems. It is a common myth in fishkeeping that large aquariums are easier to maintain than small ones; in large volumes of water pollution just takes a little longer, but when it arrives the headaches are bigger!

Even so, the realities of keeping large cichlids should and will not detract aquarists from accepting the challenge they present. The brilliance of a male *Cichlasoma* in its full display colour and posture is the making of any large aquarium.

Water requirements and filtration

Once it may have been enough to state 'bright and clean'; now water requirements for specialist groups of fishes are taken quite seriously among dedicated fishkeepers. Providing water conditions that mirror those of the natural environment not only helps to ensure the fishes' survival in captivity but is also particularly critical for successful breeding. In parallel with this greater awareness of water requirements there has been a significant increase in the sophistication of monitoring and treatment systems widely available to the hobbyist.

Here we look at the quality of water suitable for Central American cichlids, principally from the chemical point of view. We will return to the more aesthetic aspects of recreating the fishes' natural environment in the section on aquascaping, starting on page 22.

Hardness and pH value

The rivers and crater lakes of Central America are generally hard and alkaline, although some small creeks are acidic and can be more so in the dry season. 'Hard and alkaline' means a hardness of 10-30°DH (which is equivalent to 150-550mg/litre $CaCO_3$ and is described as 'slightly hard' to 'very hard' in comparative terms) and an average pH value of 8.0.

For the fishkeeper, the difficulty of providing these ideal conditions depends on the nature of the local domestic water supply. If you live in a soft water area – usually where reservoir systems provide the bulk of the domestic water supply – your tapwater may be slightly acidic and very soft. A typical example would be a pH value of 6.8 and a total hardness of 0-3°DH (equivalent to 0-50mg/litre $CaCO_3$). To raise the pH value to the required range of 7.5-8.1 and increase water hardness at the same time you can add crushed cockleshells or crushed coral to the substrate in the tank. The exact amounts required depend on the nature of the water supply. As an example, let us consider a suitably furnished

aquarium measuring 90×38×30cm (36×15×12in) containing approximately 96 litres (21 gallons) of water with a pH value of 6.9-7.1 and a total hardness of 5-10°DH. Such a system would need about 2.3kg (5lb) of crushed cockleshell added to the substrate to lift the pH value to 7.5-8.1 and approximately 30gm (just over one ounce) of calcium sulphate and magnesium sulphate to increase the hardness to 20-25°DH (330-420mg/litre $CaCO_3$).

The effectiveness of using dolomite chips in the substrate or filter body is continually under review among fishkeepers. Since dolomite – basically calcium and magnesium carbonate – is a hard, fairly insoluble substance, many fishkeepers argue that only dolomite dust can act as a useful pH buffer. Crushed cockleshell or crushed coral may be the better choice for consistent results.

It may be that your tapwater has an ideal pH value for keeping

Central American cichlids. In fact, many water authorities treat neutral tapwater so that it becomes moderately alkaline, i.e. with a pH value of 7.5 to 8.0, to reduce its corrosive effect on water mains.

Whatever type of water you have, always keep a constant check on the pH value and hardness of the water in your tank and monitor the tapwater regularly; both can fluctuate more widely than you might imagine.

The importance of water changes
Maintaining lively 'bright water' with a suitable pH value demands regular water changes. Most aquarists are aware that partial water changes

Left: *A chalk barrier in Mexico, home water to several species, including* Cichlasoma intermedium, *a relative newcomer to aquarium circles. Good water flow can be essential in aquariums to simulate the habitat.*

Below: *In keeping with all Central American riverine cichlids,* Cichlasoma intermedium *will thrive in bright alkaline water of pH 7.5-8.1.*

help to keep the system healthy, but they do not realise the importance of extracting sediment out of the substrate at the same time.

Experience shows that simple partial water changes often have no lasting effect on raising the pH value of tank water that has become undesirably low. For example, in an established tank with a pH value of 6.8, performing a 30 percent water change using a domestic water supply of pH 7.2 will still leave the tank pH value at around 6.8 or 6.9. Adding a pH corrector, such as sodium carbonate, will temporarily lift the pH value, but this effect will not last. However, extracting silt from the gravel and making a 30 percent water change at the same time, plus the addition of a pH corrector, will easily lift the pH value to the ideal region of 7.5-8.1.

Many fishkeepers are not aware that the pH value of water in the enclosed system of a tropical aquarium is influenced by many factors, including the substrate, the amount of organic debris and the number of fishes the tank contains. More fishes equals more waste and excess protein in the system, which leads to decreasing oxygen levels, a rise in carbon dioxide and a consequent fall in pH value (i.e. the water tends to become more acidic). The fishes show their distress by increased gill rates in an attempt to extract oxygen from the water in the aquarium.

Monitoring the pH value of the water regularly will reveal such trends, but more direct visual clues will help you to judge the water quality. Look for bubbles clustering on the water surface as a telltale sign of excess protein, which increases surface tension. Draw off a glass of tank water and stand it in front of a white card; if it is clearly yellow in colour then consider making a partial water change and cleaning the substrate of organic debris, as described on page 31.

Achieving a stable pH is the long-term target. Aim for it by monitoring the water regularly, by avoiding overstocking and by being generally aware of the system in your care.

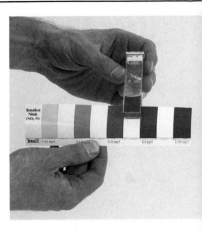

Above: *Monitoring nitrite levels in aquariums, especially new ones, is vital. Here, a sample of tank water treated with a reagent is compared to a nitrite level colour chart. Such test kits are available for checking pH value and other parameters.*

Ammonia, nitrite and nitrate

The major consideration for newly established Central American cichlid aquariums is that ammonia is a greater poison in alkaline water – i.e. within the range of pH values that are ideal for these fishes – than it is in water of a lower pH value. This factor alone has caused more fish deaths than any published statistics will ever show. (Acidic water contains a higher concentration of hydrogen ions (H+) than alkaline water and these combine with the toxic ammonia gas (NH_3) to form less harmful ammonium ions (NH_4^+). In water with a pH value of 8.0, however, five percent of the ammonia/ammonium content is present in the toxic ammonia form; at pH9 the level of free ammonia increases five fold.)

In the normal course of events, the first few weeks of a new aquarium's life is critical. If the filtration system has not been seeded with gravel or filter medium taken from a mature system (see page 19), the prospect of ammonia and then nitrite poisoning is an ever-present danger. Ammonia, a direct result of the fishes' waste products, peaks first. As the activity of nitrifying bacteria increases, much of the ammonia is oxidized to nitrite (NO_2^-), which produces a second peak. The natural process of nitrification – a process encouraged in filtration systems, especially undergravel (biological) filters – converts much of the nitrite to the less harmful nitrate (NO_3^-).

Once a system has been established for several months, ammonia and nitrite levels should no longer pose a threat. Cleaning the filtration media and making regular partial water changes should keep the nitrate levels within long-term acceptable limits.

Still or moving water?

Ideally, an aquarium housing Central American cichlids should provide both still and moving water, allowing the occupants an element of choice. In the wild, a single widespread species may well be recorded in both types of habitat. For the fishkeeper, it is easier to establish such distinct regions in a large aquarium rather than in a small

Below: *This graph shows the chemical changes that occur in the early life of a new aquarium. Within two or three weeks of fishes being introduced, the level of ammonia (A) in the water peaks. As this falls, the nitrite level (B) reaches a maximum within three to five weeks. The level of nitrate (C) builds up over several months and is reduced by regular partial water changes. Seeding new filters with mature media reduces these chemical fluctuations.*

one. As shown in the aquascape scheme illustrated on page 26, one possible strategy is to install a power filter or power head at one end of the aquarium and a gentle aerated uplift at the other. A flow would be generated around the aquarium but a pocket of relatively 'quiet' water would also develop.

Ideal though this sounds, it could lead to unexpected 'social' problems within the aquarium. Strong fishes or those at the top of the pecking order may bully those fishes that prefer to remain in the slow water areas. Breeding pairs, for example, would invariably choose the shelter of such slow water in which to spawn. Thus, in defending such ideal territory, unfortunate low pecking order fishes would be an easy target for excessive violence.

Filtration

Large healthy fishes fed correctly produce plenty of waste. In the wild, waste products are diluted immediately in relatively large (and usually moving) volumes of water. And the natural process of nitrification (i.e. the conversion of toxic ammonium compounds to less harmful nitrates by bacterial action) continually purifies the water. In the closed environment of an aquarium, however, filtration and aeration are necessary parts of the basic life support system for the fishes. To spend a higher percentage of the total budget on aquarium decoration than on filtration would be a serious error.

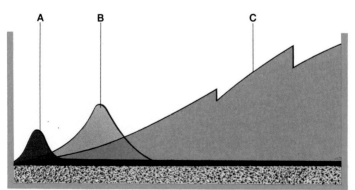

A   B   C

The ideal filtration system

The ideal filtration system for Central American cichlids is a combination of efficient undergravel filtration and external power filtration. The bacterial colony that builds up in the substrate carries out the natural process of nitrification and the external filter extracts sediment and waste from the water before it can be dragged down into the gravel. It is also possible to encourage a back-up nitrifying action in the power filters, as we shall see later.

Some fishkeepers advocate using reverse-flow filtration, a combined system in which a power filter cleans the water before it passes *upwards* through the filter plate and then through the gravel. This system is particularly beneficial for marine fishes, but for cichlids experience shows that using conventional undergravel and power filtration in parallel provides the best of both worlds, one system acting as an independent back-up to the other.

It is vital to use at least 7.5cm (3in) of rounded gravel over the filter plates. A particle size of 5mm (³⁄₁₆in) is ideal. Place rocks around the base of the uplifts to prevent them from being displaced by digging or fighting cichlids. Also, place a gravel tidy (simply a plastic

Below: *This combination of undergravel and power filtration is ideal for Central American cichlids. Each system works independently. A venturi adds beneficial air to the water outlet from the power filter.*

mesh) in the substrate to prevent digging cichlids from uncovering the filter plate.

You can fit power heads (electrically driven water pumps) to the top of the uplift tubes to speed the flow of water through the undergravel filter. Some power heads aerate the water only if they are positioned close to the water surface, so cut down the uplift tubes if necessary to produce this beneficial aeration. Most power filters have venturi adaptors to allow air to bleed into the water flow and this also enhances the level of oxygen in the water.

Although undergravel and power filtration make an ideal combination, it is possible to use either in isolation. In this case, however, you must be even more rigorous in your maintenance routine. For aquariums operating purely on undergravel filtration, be sure to remove sediment from the gravel during weekly or fortnightly partial water changes. If you aim to operate an aquarium solely on power filtration, ensure that you use extremely powerful canister filters and that the substrate in the aquarium is shallow. A light scattering of river sand or a shallow pebble layer would be ideal, but you will need to rake this through at least once a week.

Since aquascapes for Central American cichlids feature plenty of rocks and wood, and considering the large fishes possibly involved, you should seriously consider using both types of filtration in any elaborate aquarium.

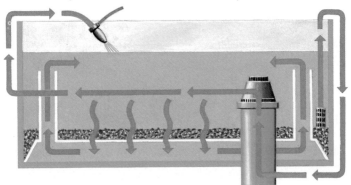

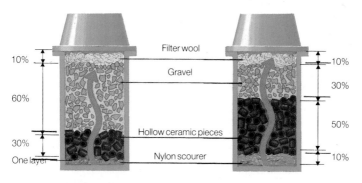

| Small/Medium cichlids | Large cichlids |

Power filter media

In order to provide additional and reliable nitrification over a long period, load power filters with long-term filter media. The alternatives of filter carbon and sponge medium will both support nitrifying bacteria but need more frequent changing. Filter carbon has been advocated for many years but its main disadvantage is that its lifespan cannot be predicted. It has an incredible surface area for adsorbing waste materials, but once this is saturated organic waste can leak back into the system. Sponge medium is more or less useless because once the surface area becomes greased with organic slime, water tends to flow over it rather than through it.

Ideally, layer the inlet/base area of the power filter with a nylon scourer and some ceramic hollow pots. This will allow water to pass through virtually unimpeded but will trap larger particles and organic debris. Fill the upper part of the filter canister with porous gravel as used in the aquarium. This will act both as a physical and as a biological filter once bacteria become established on the surface of the gravel particles. Finally, fill the upper part of the filter body with a 5-7.5cm (2-3in) layer of filter wool to extract fine particles of sediment from the water as it leaves the filter.

If the aquarium contains large cichlids, reduce the amount of gravel substrate and increase the proportion of ceramic pots in the

Above: *Recommended media for power filter canisters to ensure good water flow and efficient filtration. Use a coarser gravel for large cichlids.*

filter body. And, if necessary, increase the water flow by using a coarser gravel.

When installing a new filter, always 'seed' the nitrification process by adding some mature gravel or filter medium (i.e. used in an aquarium over 8-10 weeks old) to the new medium. A good aquarium dealer will help you with this if you cannot provide suitable material from an existing aquarium.

Selecting the correct filter size

It is impossible to over-filter aquarium water; clean water is to fishes as clean air is to humans.

For any given aquarium capacity a power filter should be capable of turning over that volume two to three times per hour *under pressure*. Manufacturers tend to quote turnover figures for their filters based on an unloaded canister under no pressure. Once sediment is drawn into the filter medium, the flow rate can fall drastically.

If your budget cannot stretch to a power filter when you set up the aquarium, then all is not lost. An undergravel filter (once it has established the bacterial colony) can cope with a new aquarium system for 8 to 12 weeks. To spread the costs, you can add a power filter at this stage without causing undue harm to your fishes.

Lighting and heating

Lighting an aquarium successfully can create a marvellous picture. On the other hand, a badly lit aquarium appears dull and unimaginative. Choosing the appropriate method of lighting an aquarium is thus an important decision for both the fishes and onlookers alike.

Fluorescent or tungsten?

Fluorescent tubes are the norm for aquarium lighting. They are cheap to run and cool in use, but they direct an even strip of light into the aquarium that produces a predictable effect. The alternative is to use tungsten spotlighting. This can create truly dramatic effects but has a number of technical disadvantages. The main objection to spotlighting – and probably the reason why it is so little used – is the difficulty of accommodating the lamp fittings into a narrow aquarium canopy.

Which system of lighting should you use in an aquarium containing Central American cichlids? Using both types in combination – an option favoured in West Germany and Holland – is an excellent choice,

Below: *This 160×60×75cm (63×24×30in) aquarium is lit by two 65 watt Grolux fluorescents at the front and two 100 watt spotlights.*

especially when such combinations include fluorescent tubes that produce a balanced light output for encouraging plants to grow in aquarium conditions. Since plant growth represents a very low priority in Central American cichlid systems, however, why not experiment with spotlamps to achieve dramatic lighting effects? And since these effects are for your enjoyment as much as for the fishes' benefit, time them to coincide with your fish-watching activities.

The best way to overcome the problems posed by the canopy is to dispense with the canopy altogether and mount spotlights above an open aquarium. Since the top of the tank invariably looks untidy, mount the aquarium a little higher than usual so that the top is not in view. You can then suspend the spotlights in attractive fittings or conceal them behind a facade. Once installed, powerful spotlights directed at a rippling water surface will produce a striking display of angled shafts of light in the aquarium. The cichlids will quite happily swim in and out of these lighted areas.

Whatever type of lighting you use, be sure to keep the aquarium regularly maintained so that the lighting is used to its maximum

effect. An excess of tannic acid in the water – leaching from unsoaked bogwood, for example – will cause a yellow-brown cast that severely reduces the penetration of light. Efficient filtration and regular partial water changes should prevent any major discoloration of the water. It is also vital to keep the cover glass clean to allow as much light as possible to reach the aquarium.

Heating the aquarium
What temperature suits Central American cichlids? In the wild, they seem able to thrive in a wide range of temperatures. Hans Mayland, a West German aquarist, has collected cichlids in various countries. In the Panamanian rivers of Capira, Mendoza, Chagres and Tupisa he found cichlids flourishing in recorded water temperatures of 25.6-31.2°C (78.1-88.2°F). (Air temperatures were in the range 22-33°C/72-91°F.)

The fact that cichlids can survive this temperature range in the open ecosystem of their natural habitat does not mean that they are as tolerant in captivity. In the closed environment of an aquarium, excessively high temperatures may lead to unacceptably low oxygen levels in the water and produce high gill rates and stress among the fishes. With efficient filtration and aeration, however, it should be possible to maintain the water temperature at 27-29°C (80-84°F). The advantages of keeping the water at these fairly high temperatures are that the fishes are more active, consume more food and thus grow more quickly than in cooler conditions. On the debit side, bacterial infections are more rapid at higher temperatures and parasitic infestations cycle at a faster rate. (Although, perversely, this may make parasites easier to treat.)

As far as hardware is concerned, use the standard heater-thermostats that are universally available and reliable in use. The table shows the recommended number and power ratings for various aquarium sizes. If in doubt, ask your local dealer for advice.

Recommended lighting systems

Aquarium size (L×D×W)

60×38×30cm (24×15×12in)
1×45cm (18in) Grolux (15 watt)
12 hours per day
or
1×60 watt spotlight 6 hours per day

90×45×30cm (36×18×12in)
1×75cm (30in) Grolux (25 watt)
12 hours per day
or
1×75 watt spotlight 6 hours per day

120×45×38cm (48×18×15in)
2×90cm (36in) Grolux (30 watt)
12 hours per day
or
2×75 watt spotlights 6 hours per day
or
1×90cm (36in) Grolux (30 watt)
12 hours per day combined with
1×75 watt spotlight 6 hours per day

150×45×38cm (60×18×15in)
2×120cm (48in) Grolux (40 watt)
12 hours per day
or
3×75 watt spotlights 6 hours per day
or
1×120cm (48in) Grolux (40 watt)
12 hours per day combined with
2×75 watt spotlights 6 hours per day

Recommended heating systems

Aquarium size (L×D×W)

60×38×30cm (24×15×12in)
1×200 watt heater-thermostat

90×45×30cm (36×18×12in)
2×150 watt heater-thermostats

120×45×38cm (48×18×15in)
2×200 watt heater-thermostats

150×45×38cm (60×18×15in)
3×200 watt heater-thermostats

(The recommended heater ratings are slightly higher than are normally suggested, but experience shows that this provides a vital safety margin.)

Aquascaping

Aquascaping an aquarium is important for two fundamental reasons: to provide a stable and easily maintained environment in which the fishes can thrive and to produce an aesthetically pleasing 'picture' for onlookers to enjoy.

Recreating the fishes' natural habitat is one of the greatest challenges that faces fishkeepers and is often the priority for all animal keepers. Working within reasonable parameters, it is possible to provide a cichlid community with an environment which, although not quite in the natural dimension, parallels the real situation sufficiently to sustain the aquarium occupants. The beauty of an aquascaped aquarium is very much in the eye of the beholder, but most people – even non-fishkeepers – would agree that a successful result can enhance a living room and provide an endless source of conversation.

Behind the 'scenes' of any aquarium layout lie the essential services of filtration, aeration, lighting and heating, which we have considered in their own right in earlier sections. Here we concentrate on the environmental and artistic elements of aquascaping and look at three recommended schemes as examples of effective aquascaping in action.

The natural environment
The narrow land bridge of Central America from Mexico through Guatemala, Belize, Honduras, El Salvador, Nicaragua, Costa Rica, Panama to Colombia in South America is veined with rivers and marked with small lakes. These waterways provide a wide range of environments; some small tributaries of major rivers are fast flowing, others are idle creeks. The flowing water, unless in full torrent, is usually clear, often agitated as it passes over rock rubble and small boulders. The lakes vary in depth, sometimes according to season. Several styles of aquarium aquascapes, therefore, are needed to imitate these natural habitats.

Selecting the correct substrate
The substrate dictates the overall atmosphere of any aquarium. In the wild, Central American cichlids swim above pebbles, leaf-littered mud or sand. Large pebble gravel can look fantastic in an aquarium but allows uneaten food to become trapped and pollute the water. River sand, superb when rippled by a water flow, can become packed and starved of oxygen. This can lead to adverse bacterial activity and serious infections spreading among the fishes in the aquarium.

As we have seen on page 18, external power filtration can maintain the water quality in aquariums using either of these

Below: *This fast-flowing highland river in Costa Rica illustrates the rock rubble habitat that an ideal aquarium aquascape should simulate.*

Above: *This aquarium uses plastic plants and rounded boulders to good effect. Combining real and plastic plants also looks very natural.*

substrates. If undergravel and power filters are operated in conjunction, both correctly established and maintained, then pebble gravel or river sand are ideal and safe aquarium substrates for Central American cichlids. Naturally, the 'standard' substrate of rounded gravel of 5mm (3/16in) particle size provides a reliable foundation for any cichlid aquascape.

The boundaries of a fish's territory may be closely linked to the pattern of the substrate. A cluster of four or five small boulders, for example, may form the dividing line between one territory and the next. In a crater lake, a piece of driftwood may act as a territorial 'marker'; in a river it could well be a cobbled area or a fallen tree. Territorial boundaries also depend on the size of the cichlids in question and the number of fishes in a population.

Planting
When it comes to discussing plants and large cichlids, most fishkeepers throw up their hands in despair; the two are often thought to be incompatible. Certainly, dainty plants are very quickly devoured by cichlids on the rampage or in search of food. Nonetheless, some species

of plants are sturdy enough to survive the rigours of a Central American cichlid system.

The evergreen Java Fern (*Microsorium pteropus*), *Hygrophila corymbosa*, the False Cryptocoryne (*Spathiphyllum wallisii*) and the Amazon Sword (*Echinodorus* sp.) are all tough enough to consider planting in a well-fed cichlid system with good water conditions. Java Fern is a particularly tough plant. It will thrive in alkaline, even brackish, water and would be difficult for even the largest cichlid to consume.

Combining suitable real and artificial plants can also be an extremely effective strategy. The art of using artificial plants successfully in a living aquarium relies on creating the impression that the plants are growing. Grouping the plants in clumps produces a convincingly natural effect. A 30-38cm (12-15in) plastic *Vallisneria*, for example, looks more 'alive' with two or three 15cm (6in) versions arranged in a cluster around the base. After a few weeks, algae forms on the synthetic fronds and the illusion is complete.

Recommended layouts
The three alternative aquascapes featured on the following pages strive to satisfy the needs of aquarium fishes in functional and attractive simulations of their natural habitats.

Aquascape 1

Based on an aquarium measuring 60×38×30cm (24×15×12in) or 90×45×30cm (36×18×12in), this layout is ideal for keeping small cichlids, such as *Cichlasoma spilurum* or *Neetroplus nematopus*. It would be an eminently suitable layout for fishkeepers new to Central American cichlids. Use a basic substrate of rounded gravel with a particle size of approximately 5mm (³/₁₆in) and to a minimum depth of 7.5cm (3in).

The main 'architectural' feature of the layout consists of a cluster of rocks and slate. Build this up at one end of the aquarium using water-worn rocks from hill streams or hard, sea-washed (and suitably rinsed) rocks from the beach. It is important not to make the cluster too high because spawning cichlids will dig out gravel at the base and may cause the stones to topple and fall against the aquarium glass. You can prevent cichlids digging too far down into the gravel by placing a sheet of gravel tidy about 7.5cm (3in) below the surface of the gravel. Nevertheless, you can insure against the possibility of an underwater rock fall by looking for a single piece of well-jointed rockwork to form the centrepiece of the arrangement. Finally, butt up pieces of slate to the rounded boulders to create a series of 'natural' crevices the fishes can use.

Litter a few pebbles in the centre of the aquarium floor for the fishes to 'associate' around and decorate this area with a real Sword Plant (*Echinodorus* sp.) or its plastic equivalent. Arrange some pebbles at the opposite end of the tank, perhaps more than the token amount used in the centre, and frame an upturned or side-down plant pot with living or artificial *Vallisneria* plants. A small piece of bogwood completes the scene.

In a 60×38×30cm (24×15×12in) aquarium you will need only one undergravel filter plate and its associated air-operated uplift. For a 90×45×30cm (36×18×12in) tank, use two equally sized filter plates; position the uplifts at opposite ends to enhance the flow of water around

Use tough plants, such as Swords or *Vallisneria*, or their plastic equivalents.

Above: *An aquascape based on a 60×38×30cm (24×15×12in) tank.*

Right: *A spawning pair of* Cichlasoma spilurum; *the larger fish is the male. This species is well suited to medium-sized aquariums, and such a pair could successfully raise fry in the aquascaped tank shown above.*

the aquarium. If you use power heads at the top of the uplifts, you can position these side by side in the centre back of the aquarium and conceal the whole arrangement with a single piece of bogwood. The powerful flow of water produced by the power heads will be more than sufficient to distribute the heat from the heater-thermostat.

A half-buried pot provides an ideal refuge or breeding site for territorial fishes.

Rounded boulders and a piece of slate create a natural series of crevices for the fishes to use.

Bogwood – an essential part of any Central American system.

Aquascape 2

This layout allows more swimming space for larger cichlids, such as *Cichlasoma nicaraguense*, *C.salvini* and *C.synspilum*. It is based on an aquarium measuring 120×45×30cm (48×18×12in) or 120×45×38cm (48×18×15in). The focal point of the design is an arrangement of rounded boulders framed by a large overhanging piece of bogwood. If you cannot find a suitable single piece, build up the desired shape by gluing several smaller fragments together with aquarium sealant.

To one side of this structure build up a 'cavework' with small boulders and a roof of slate. This will provide territory for a single pair of cichlids. If a group of juvenile cichlids are

Left: *Cichlasoma synspilum*, a colourful and widely available species that would be an excellent choice for the aquascaped tank shown below.

Plastic plants arranged in natural clumps can look very realistic.

A gentle uplift creates an area of quiet water for subdominant fishes.

continually fighting to move up the pecking order or to establish territory, make up an extra cluster of rockwork at the other end of the tank. In most cases, it is more preferable to provide a surplus of shelter/territorial options rather than just a sufficient number. The open areas left in the aquarium will allow the cichlids to spawn, perform and generally posture against each other as they do in the wild.

Use a gravel substrate as specified in the first aquascape. Fit two undergravel filter plates in the aquarium and operate them with a simple air lift and power head respectively. This will provide a beneficial combination of water flow and aeration around the aquarium.

Below: A substantial piece of bogwood lies at the heart of this aquascape for a 120cm (48in) tank. The result is spacious but provides plenty of territorial refuges.

Use a power head at one end of the aquarium to create beneficial water circulation.

Plenty of crevices here for the fishes to use either as territorial refuges or breeding sites. Such cavework features are easily built up from rocks and pieces of slate.

Scatter large smooth pebbles in groups to form 'association' areas for the fishes.

The best substrate to use over an undergravel filter is rounded gravel of 5mm (3/16in) particle size, at least 7.5cm (3in) deep. Use a gravel tidy to prevent digging cichlids uncovering the plastic filter plate.

Use a well-shaped piece of bogwood as the central feature in the design. Wash it before use to avoid leaching.

Aquascape 3

A few specimens of the larger species, such as *Cichlasoma motaguense* or *C.managuense*, or a group of the smaller species would find this expansive aquascape a true 'home from home'. The layout is based on an aquarium measuring 150×45×38cm (60×18×15in). At first sight it seems cluttered. The combination of boulders, bogwood and general litter, however, creates an environment in which the strong can form territories and the weak can escape the attentions of bullies.

Use three undergravel filter plates in the base of the aquarium and cover these with at least 7.5cm (3in) of rounded gravel as recommended for the other aquascapes. Place a large piece of bogwood in one corner of the aquarium and arrange a beechwood branch so that it extends into the centre foreground. Along the back of the tank build up a wall of large boulders, taking care to make sure they are stable. In the opposite corner, arrange rounded boulders and pieces of slate to

Arrange large boulders and slates to form a natural series of caves.

Ensure that the rocks are stable. Large cichlids can easily dislodge stones and cause damage to the aquarium glass.

'Drive' the middle undergravel filter plate with a simple air uplift system.

Use a high-capacity power head here to create a strong water flow towards the air uplift situated at the centre back of the aquarium.

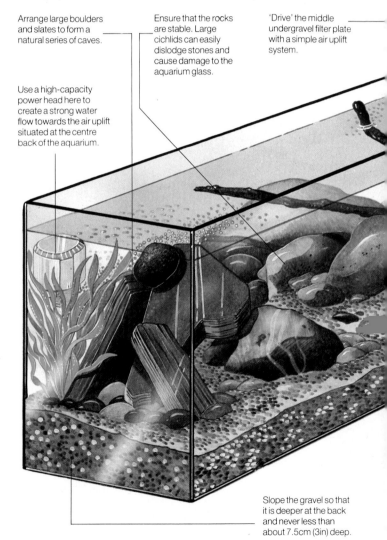

Slope the gravel so that it is deeper at the back and never less than about 7.5cm (3in) deep.

create crevices, caves and holes. Position plastic plants (Sword Plants are ideal) here and there to complete the scene.

Using a combination of air-operated and power head uplifts for the undergravel filters satisfies the twin demands of aeration and good water flow necessary in such a large aquarium. If you direct the outlet of the power head located behind the cave area across the back of the aquarium towards the airlift tube in the centre, this will drive air bubbles around the aquarium and create a water flow towards the power head in the opposite corner. This second power head will move water towards the front glass and set up a beneficial circulation around the entire aquarium. You will find that the fishes will often swim around, along or into the powerful streams of water issuing from the electrically driven power heads.

A beechwood branch and large piece of bogwood create a focal point in the design. Behind the bogwood use a power head to direct water to the front.

Above: A large varied aquascape for 150cm (60in) aquariums, allowing plenty of swimming room.

Right: An adult *Cichlasoma managuense*. Large specimens need space and can readily dominate their tanks.

Feeding and routine maintenance

The high quality and range of prepared and frozen foods now available means that fishes in captivity are probably in a better condition than those fending for themselves in the wild. And if water quality is maintained many, if not all, fishes live longer in aquariums than in their natural habitats. In the aquarium, they are protected from droughts and floods and have food placed before them every day! In the wild, food is not always constantly available; because of season or circumstance some days it may be plentiful, some weeks it may be scarce. To echo this natural irregularity, consider leaving the fish unfed on one day each week. Although this strategy is open to debate, there are sound reasons behind it. In most cases, aquarium fishes are overfed. Thus, missing a day tends to take the pressure off the system in terms of dealing with organic waste. If you have spawning or brooding fishes in your aquarium, however, do not interrupt the feeding routine. Parent fishes take in food and blow out any excess through the gills; this is an important source of food for fry (see page 38).

Below: *A pair of* Cichlasoma bifasciatum *feeds on lettuce. A varied diet is essential for Central American cichlids and some need regular amounts of green food.*

What do they eat?

In their natural habitat, Central American cichlids eat plant debris, young fishes, crustaceans, insect larvae, and terrestrial insects and worms. In the aquarium, these fishes will readily accept prepared foods, such as large flake foods, pellets and foodsticks, as the bulk of their diet. Feeding a selection of frozen and live foods on a rota basis, such as brine shrimp, *Mysis* shrimp, earthworms and bloodworms, will keep the fishes in excellent health.

The ideal feeding strategy

Variety is the key factor for successful feeding. Offer established fish communities several feeds during the daytime or evening and alternate the type of food. On any one day, for example, offer flake, pellet or foodstick, and frozen or live foods in a planned rota. Some species, such as *C.sieboldii* and *Neetroplus nematopus*, are known to eat algae and other plants in the wild. An aquarium equivalent of this natural diet would be leaf spinach, lettuce and certain vegetable flake foods.

Try to avoid using the powdery remnants of flake food when the tub is almost empty. The fishes refuse these small particles and they pollute the system. If possible, buy the 'large flake' brands or 'cichlid flakes' to avoid this problem.

Above: *Using a siphonic cleaning device to extract organic debris from between the gravel particles. Clean a small section of substrate at a time.*

As with all aquarium fishes, however, resist the temptation to overfeed Central American cichlids; give the fishes no more food than they can consume in a few minutes.

Overfeeding or a breakdown in the bacterial balance of the system can sometimes show itself by the presence of planarians in the gravel and on the glass. These free-living flatworms do not appear to harm fishes (although they will destroy eggs and fry) but their proliferation usually indicates an excess of food in the aquarium. Use a copper-based or general anti-parasite treatment and reduce the amount of food for a period of several days to discourage the planarians. Do not make large water changes with raw tapwater, however, because this will suppress the natural nitrification processes carried out by bacteria in the undergravel filter. If you do feel a water change will help the situation, aerate stored quantities of water overnight to help dissipate the purification additives.

Routine maintenance tasks
In the sections on water requirements (pages 14-17) and filtration (pages 17-19) we have considered the vital role that water changes and the selection of filters play in controlling pH balance and establishing a stable aquarium environment for Central American cichlids. Here, we review the routine maintenance tasks common to all aquarium systems.

Perhaps the most basic routine check to make is a 'roll call' of the fishes in your aquarium. A large body concealed behind a rock or piece of bogwood could pollute the aquarium if you leave it to decompose over a period of time.

Maintaining filtration systems
Regular maintenance is one of the most important tasks for the conscientious fishkeeper and is especially vital for filtration systems.

Every week check the power head or power filter flow for strength. A reduced flow on a power head signals a blockage in the undergravel filtration system and the time may have arrived to clean the gravel. Use one of the inexpensive syphonic action devices to draw up debris from between the gravel particles. Since the organic debris is lighter than the gravel, it is whisked away in the syphonic flow of water set up in the funnel while the gravel remains in the aquarium. Cleaning the gravel in this way every three or four weeks will prevent accumulating dirt from disturbing the pH balance of the water and ensure the free passage of oxygen so essential for the nitrifying bacteria to flourish. It is the equivalent of a gardener turning over the soil.

Cleaning power filters is also a regular but less frequent task. Leave power filters running until the flow rate is clearly reduced by 60-70 percent. This indicates a significant build up of sediment inside the canister. To clean the filter, empty all the contents into a bucket of water taken from the aquarium, rinse the medium and filter wool thoroughly and return them to the canister. The important point here is the use of aquarium water to clean the filter contents. Avoid the temptation to wash the filter medium under running tapwater; the chlorine in the water can suppress the bacterial activity established in the filter. Rinsing the medium in aquarium water of the correct pH and temperature does not disturb this bacterial activity.

Inspect airstones used in undergravel filter uplifts. A noticeable reduction in aeration usually points to the deterioration of an airstone. These accessories are cheap to replace and if left on the filtration system when blocked they can reduce the efficiency of both filtration and pump. A blocked airstone can put unnecessary back pressure on the air pump, for example, and this in turn will shorten the life of valves and diaphragms.

Checking pH and temperature
Take pH and temperature readings every week. A falling pH (i.e. tending towards acidic conditions) usually indicates a need to extract sediment from the substrate and power filter media as described above. Normally, regular water changes combined with the removal of sediment will prevent serious changes in the pH value of the water, as we discussed on page 16. Overstocking and excessive feeding of large fishes will also cause a falling pH value.

Water temperature can range widely if the room temperatures are extreme. In normal circumstances, large temperature fluctuations point to the failure or inefficient operation of the heater-thermostat or that its power rating is inappropriate for the intended use (see page 21). To guard against excessive cooling, some fishkeepers install a back-up heater-thermostat set to a lower temperature than the main unit or units. The majority of temperature regulation problems seem to cause the opposite effect, however. Experience shows that the most common fault is the failure of the thermostat to cut out the heating element. The subsequent rise in temperature usually reduces the oxygen levels in the water to unacceptable levels and fishes are lost as a result. Ideally, you should get into the habit of placing your

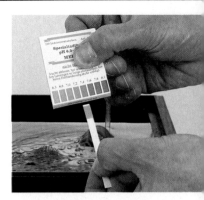

Above: *This paper strip indicator provides a quick way of taking a pH value reading. Monitor tapwater and aquarium water pH values regularly.*

hand on the aquarium glass to gauge the relative warmth of the water, especially last thing at night or early in the morning. If the tank feels warm relative to your own body temperature of approximately 37°C (98.6°F) then check the thermometer immediately.

Testing nitrite levels
It is advisable to test for nitrite levels during the first six weeks of an aquarium's life – while it is still maturing in the bacterial sense – or if you suspect that the filtration system has broken down. Such breakdown might show itself in terms of cloudy water and/or the fishes showing signs of itching. If an excessive nitrite peak occurs during the maturing process (a transient peak is natural), try introducing gravel from a mature system or use a proprietary additive designed to enhance the nitrifying cycle. Once such a peak is over in a new system, it rarely recurs.

If nitrite levels remain unnaturally high in an established system, discuss the matter with your dealer, who should be able to offer valuable insights into the problem related to your particular set-up. The problem may arise because the natural bacterial activity has become disturbed by over-zealous water changes or by the use of raw tapwater to wash filters.

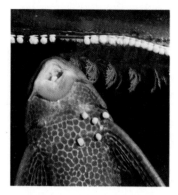

Above: *A large Suckermouth Catfish* (Pterygoplichthys *sp.*) *takes excess cichlid pellets from the surface. An ideal scavenger for a cichlid tank.*

Coping with algae

If algae forms on the front glass, scrape it off using a nylon or plastic pan scourer, but do not use the type impregnated with soap or other cleaning agent. Cover glasses are also prone to algae coating; the damp, well-lit conditions are ideal for the growth of these plant cells. If you keep the algae at bay, cleaning all types of glass or plastic condensation covers should not prove too difficult. But algae that has built up over a long period can be difficult to remove.

Algae on rocks can be considered natural and a beneficial influence in balancing the aquarium system. Algal cells absorb nitrate from the water as a food source (hence nitrate fertilizers) and provide a large surface area to support the oxygen-fed bacteria that 'drive' the nitrifying cycle. And algal growths can form a useful food source for certain cichlids; *Neetroplus nematopus* is one of the cichlids that has evolved teeth structures to enable it to browse on algae.

Rampant algae, however – in some varieties filamentous algae – can be unattractive and harmful to the system. An excess of algae usually shows that the lighting is too strong or has been left on too long. Direct sunlight striking the tank at certain times of the year may also cause algae problems. Sunlight creates a pleasing effect and should not be avoided completely, but if it causes unsightly 'algal blooms' across the substrate and glass, then you must take steps to control the amount reaching the tank.

Help with the 'housework'

Large cichlids produce copious amounts of waste and are inefficient feeders, especially of prepared foods such as pellets or foodsticks. When eating these foods, cichlids spit or blow out clouds of food particles. This increases pressure on the 'cleaning' processes (natural or assisted) that take place within the aquarium and highlights the need to monitor water chemistry very closely and carry out regular, partial water changes.

To help with the 'housework' in a community system containing large cichlids, you may find it advisable to introduce a catfish as a substrate scavenger. Members of the armoured Catfish Family, Loricariidae, are tough enough to withstand the close aggressive attentions of cichlids. They will eat almost any type of food, including algae, but will not predate on fish eggs. The Suckermouth Catfishes of the genus *Hypostomus* (of which *H.plecostomus* is one of over 100 species known) are ideal catfishes to include in a community of large cichlids. They use their rasping teeth to scrape algae from rocks and other surfaces.

Scaleless catfishes of the genera *Rhamdia* and *Pimelodella* – both found throughout South and Central America – are also suitable for alkaline systems. Unfortunately, they *will* eat eggs and fry, presuming the brooding parents are caught off guard long enough. If spawning and rearing cichlids are priorities, then do not include *Rhamdia* or *Pimelodella* in the aquarium.

Overstocked, under-sized aquariums will always present problems. If you use a large enough tank with sufficiently powerful 'life-support' systems, regular maintenance will keep a Central American cichlid community in good condition.

Basic health care

The best way of avoiding diseases in the first place is to buy healthy fishes from a good aquarium dealer who *clearly* maintains high water quality in his aquariums. Ask penetrating questions about the requirements of the fishes *before* you buy them; a good dealer will always provide the answers you need.

Some common diseases and infections arise when the fishes are damaged or there is an equipment failure. Here we look at the cause and treatment of some of the most common health problems likely to affect your fishes.

Common health problems

External damage to the body or fins of cichlids arising from pecking order disputes is more or less cosmetic. If the aquarium is well maintained, adding a basic antibacterial treatment to the water over a period of several days will usually help the fishes recover.

If the fighting is continuous and one fish becomes the target for attack, isolate that individual from the main system to protect it from further abuse *before* using any medication to treat any possible infection of its wounds. Where flesh damage is severe, recovery will be slow, but fishes are surprisingly resilient in good conditions.

Equipment failure, such as a heater-thermostat breaking down or a power filter becoming blocked, can cause an outbreak of white spot (ich). Most experienced aquarists are not afraid of this parasitic infection. (The white spots are caused by accumulations of single-celled parasites – *Ichthyophthirius multifiliis* – living just beneath the skin.) Several effective remedies are widely available; keep one on hand in case you need it.

It is always best to treat the aquarium twice or even three times over a 10-day period because cysts may remain in the system that could cause a secondary cycle of infection. All fishes carry the possibility of developing white spot; stress, poor conditions during shipment and dramatic temperature changes can bring it on. Healthy fishes, however, can easily shrug off its effects. Ideally, always dose the aquarium with a suitable white spot treatment when you introduce new fishes after their quarantine period.

Always try to identify the cause of white spot, especially if it occurs in well-established fishes. If a heater has failed, fit a new one quickly.

So-called 'hole-in-the-head' disease (caused by the single-celled parasites *Hexamita* and *Spironucleus*) can affect cichlids in overcrowded aquarium systems or in conditions of poor water quality. A proprietary treatment is available to treat *Hexamita* and specific anti-protozoan drugs, such as metronidazole and dimetridazole (for which a veterinary prescription may be necessary) may also be effective. Some fishkeepers advocate using antibiotics to treat 'hole-in-the-head' disease. All these treatments should be administered to the affected fishes in a separate aquarium. In practice, the success rate for cichlids is rather poor once the disease manifests itself, but few other options exist.

The majority of bacterial and parasitic infections are wrongly identified by aquarists, who consequently apply an inappropriate treatment to the aquarium. The true fungus (*Saprolegnia*), for example, rarely occurs. The general white tissue damage sometimes attributed to fungus is usually caused by a bacterial infection. Always check the signs and make the correct diagnosis before taking any action. When in doubt, ask experts.

Many books write about gill flukes (*Dactylogyrus* and other species). Once a fish begins to itch and scrape its body on rocks, gill flukes are invariably suggested to be the cause. In fact, gill flukes are likely to be at the bottom of the list of possible causes. Here again, it is vital to rule out the most likely causes before blasting the aquarium with the wrong treatment. And remember, healthy fishes rarely succumb to disease if the water quality, stocking levels and feeding programme are correct.

TABLE OF DISEASES

Signs	Possible causes	Action
Fishes itching and scraping body against hard objects	White spot (Ich) (*Ichthyophthirius multifiliis*)	Treat with white spot remedy
	Excess ammonia/ nitrite levels in a new aquarium	Carry out a partial water change and seed filter bed with mature gravel
	Freshwater velvet (*Oodinium pillularis*)	Use a proprietary velvet treatment
	Low pH (3.8-5.5)	Extract organic debris from the gravel using a syphonic gravel cleaner and carry out a partial water change
	Water poisoned with insect spray, cleaning agents, etc	Carry out a 100 percent water change. Use dechlorinating chemicals to avoid damage to filter beds
	Gill flukes (*Dactylogyrus* and other sp.)	Use a 3 percent salt bath (for adults only) or a proprietary anti-parasite remedy
Fishes gasping at the water surface	Low oxygen level	Check aeration system. Check pH value
	Low pH (below 7.1)	Extract organic debris from gravel and carry out a partial water change
	Excess ammonia/nitrite levels in a new aquarium	Carry out a partial water change and seed filter bed with mature gravel
	Freshwater velvet (*Oodinium pillularis*)	Use a proprietary velvet treatment
	Water poisoned with insect spray, cleaning agents, etc	Carry out a 100 per cent water change. Use dechlorinating chemicals to avoid damage to filter beds
Cloudy eye, fin rot, etc., as signs of bacterial infection	Low or incorrect pH Poor water quality	Carry out a water change and remove sediment from gravel. Add a pH corrector. Add a general antibacterial treatment or tonic
	Inadequate filtration	Check filter and pump size to see if they are capable of coping with the system
	Overstocking	Check stocking levels
Body wounds	Stressed fish (an attacked subdominant or female fish bullied by a dominant male)	Remove to a separate aquarium and use a general antibacterial treatment
Loss of appetite and hole-like lesions in the head	*Hexamita/Spironucleus* ('Hole-in-the-head')	Use proprietary treatment or specific drugs and/or antibiotics available on a veterinary prescription

Breeding

Aquarists wishing to breed Central American cichlids must have an inner desire to attempt the difficult. Nevertheless, many species are regularly spawned and raised by dedicated enthusiasts in the USA and Europe. Other spawnings are often accidental in a community of cichlids or other tropical fishes. The high degree of parental care shown in all species can often guarantee some fry will survive even in the most intensely populated aquarium.

Some species are so large, however, that only massive aquariums can accommodate the breeding territories and the defensive roles they play. In small aquariums, a breeding pair can disrupt all other occupants and even kill a couple in the process. Such is the strength of instinct to protect the offspring that sometimes a spawning pair can be the only fishes to survive the event!

Selecting a breeding pair
Sexing most cichlids is a difficult affair, especially when they are in the juvenile or semi-adult stages. Since few breeding adult pairs can be purchased without a bank loan, juveniles are the most likely breeding stock for most

Above: *A pair of* Cichlasoma cyanoguttatum *tending fry. Once established in a system, breeding and raising fry is a natural progression.*

fishkeepers. It is fairly certain that a pair will develop among a group of juveniles, although if the selection is made from one batch, inbreeding could result by the cross of brother to sister. The best way of avoiding this is to buy the same species from different sources.

It is almost impossible to select two fishes from a batch, place them in an aquarium and expect them to accept each other and spawn. It is quite possible, however, to study a group of cichlids and observe a dominant individual and presume

Below: *Jaw locking between an Oscar (left) and a Jaguar Cichlid. This is usually a test of strength between rivals, perhaps vying for a mate.*

this to be a male. Then, a less sturdy fish with shorter fins could be selected as a possible female. If they are juveniles, it is best to isolate them in separate aquariums where each can mature. Many fishes are lost by being introduced in 'pairs'; the dominant fish almost automatically begins to attack the subdominant one. This is the problem of an enclosed system.

In their natural habitat, semi-juveniles exist in quite large shoals in relatively spacious expanses of water. It is almost impossible, therefore, for an individual to dominate another and cause its eventual demise. In captivity, the dimensions of an aquarium may simply mark out an arena in which a weak fish is killed by an aggressor. Such arenas do not occur in the natural environment; a dominant fish will simply chase off an intruder. In an aquarium, a beaten fish often has nowhere to go to escape attack.

Since good fishkeepers do not wish to witness unnecessary carnage in their aquariums, it is important not to place fishes into this situation. Always buy a group of four to six fishes and make sure you can accommodate them at their eventual size. Most aquarium

dealers will exchange well cared for cichlids that have outgrown their original accommodation.

Spawning and rearing
If you can establish a pair in an aquarium in isolation, then breeding them and raising fry is simply a natural progression.

How will you know that your fishes are spawning? When you keep cichlids for a reasonable period of time, it becomes obvious when a pair are setting up a spawning sequence. Pairs not yet sexually mature will enter into practice rituals, which usually involve defence of an area of rocks or general debris. The pair will mouth and scrape clean areas of rock or bogwood, preparing the surfaces that the eggs will adhere to. They will circle each other, the male extending his fins in much the same way as a displaying male bird. Sometimes, jaw locking will occur. It is not exactly clear why such head on tests of strength (usually between males) are important, but it probably helps to ensure that the strongest male is the selected fish to fertilize the female's eggs.

If the pair are mature enough to spawn then the female will make runs along the pre-cleaned surfaces and eventually begin to deposit groups of eggs. The male follows the female's run and fertilizes the eggs. The pair will perform this

Below: *A Convict Cichlid,* Cichlasoma nigrofasciatum, *having spawned inside this conveniently placed pot, cares for the eggs until they hatch.*

exercise repeatedly, sometimes over a long period of time, until all the eggs have been deposited. Then both fishes will fan the eggs, using their ventral and pectoral fins to sweep water over them.

At this point, immature or unsuited pairs sometimes consume the eggs or take little interest in them, thereby laying them open to predation by other fishes. On certain occasions, the pair will argue and a fight between them will end up with a badly damaged female. By contrast, a balanced well-suited pair will develop a strong bond and defend the eggs as their first priority. In this instance, any cichlids or other fishes foolish enough to encroach upon their territory will be driven off with a fervour rarely matched in the natural world.

After several days of close attention to the eggs, fry will begin to emerge in small batches until all the fertile eggs hatch. During this period, some aquarists advocate adding a small quantity of a suitable antifungal agent to the water. Use these products – which contain such substances as acriflavine, methylene blue or malachite green – according to the maker's instructions. When treating eggs, always underdose the aquarium to be on the safe side. Such treatments help to combat fungus that invariably develops on infertile eggs and which may affect fertile ones. To some extent, the parents take care of this problem by picking off any infertile eggs from the batch. Once all the fry are free of the egg cases, the parents usually place them in an area in which they can best be defended and cared for.

If the aquarium has a gravel substrate, for the next few days the fry are placed in hollows or pits dug by the parents. Within 48 hours of being free swimming, after the yolk sac has been consumed, you can feed the fry on freshly hatched brine shrimp, microworms or liquid and powdered fry foods. (The brooding parents often offer sustenance to their fry by blowing out a cloud of chewed food.) During the next 14 days offer the fry food on a 'little but often' basis.

When they have begun to forage for food away from the protection of their parents, remove the fry to a raising aquarium. Opinions differ on this strategy; some aquarists recommend leaving some of the fry with the parents so that the normal cycle of instinctive care continues in the natural way. Certainly, removing all the fry can sometimes cause aggressive disputes to break out between the parents, and the female may be seriously injured as a result. Conversely, in some instances, fry can remain with the parents too long, leading to problems developing between the

Below: *A female* Cichlasoma hartwegi *in brood colour display, caring for free-swimming fry. Parents often spit out food for their brood.*

Above: *A fry-rearing set-up suitable for Central American cichlids. Ensure that the system is well established (in the filter sense) before adding any fry. Newly hatched fry can be fed more easily in the net and, once grown, released into the tank.*

parents when another spawn cycle begins. It is not unusual for the parents to lose interest in their brood after a period of about three or four weeks, which suggests that removing the fry from the aquarium at an earlier stage is advisable.

Ensure that a raising aquarium is well established before transferring fry to it from the main tank. Use a 50/50 blend of new and main aquarium water in the nursery tank and install an internal power filter, sponge or box filter. These simple filters are ideal because they do not draw food particles away from the fry as would happen among the gravel particles of an undergravel filter. In fact, sponge filters actually provide a useful feeding surface for the fry, since they can pick off fragments of food that collect on the exposed filter body. Use these filters in a tank with a bare glass base so that any excess food can be removed easily with a siphonic cleaning device. This will prevent pollution of the water.

The larger the raising aquarium, the better the growth rate will be. Without doubt, the most important factors that affect growth rate are related to the size of the aquarium, the amount of food and the frequency of feeding. If a large raising aquarium is not practical then frequent partial water changes will help to alleviate the problems of overcrowding fry.

It is not unusual for whole batches of fry to be lost, even when everything seems to be going well. In a crowded system, any outbreak of disease spreads with truly unbelievable speed, and a sad aquarist is often unable to accept the real reasons for this.

The breeding sequence described above applies to a spawning pair of cichlids in a separate aquarium. Should a pair spawn in a community system of cichlids and other species, then the pattern is much the same. The egglaying rituals and protection of the spawn will go on despite a crowded aquarium. Developing fry will group close to their parents and will be suitably protected, but predation will undoubtedly occur once the fry begin to stray. This is a normal pattern that simply reflects the universal laws of natural selection; the strongest and most genetically ideal individuals will progress and grow.

In a community aquarium, take care to choose suitable species for inclusion with the cichlids. Many, if not all, catfishes, for example, are nocturnal and active night feeders. Cichlids will slow down their body system at night and therefore will not be capable of protecting fry just when predatory catfishes are most active. (See also page 33.)

Species section

In this part of the book we look in detail at a representative selection of 33 species of Central American cichlids, both established favourites and brilliant newcomers to the aquarium scene. They are presented in alphabetical order of scientific name, with the most widely used common names listed for each species. The majority fall within the genus *Cichlasoma*, and for simplicity we have elected to keep them there. As with so many branches of natural science, moves are continually underway to reclassify fishes into different taxonomic groups. Such moves have become particularly strong as far as Central American cichlids of the genus *Cichlasoma* are concerned. Revisions of scientific nomenclature are the production of detailed study at the highest academic levels. Since the concerns uppermost in the minds of professional ichthyologists do not necessarily coincide with the practical considerations of amateur fishkeepers, however, we have resisted

the temptation to enter the nomenclature fray until a clear picture has emerged. In the interim, therefore, we have not introduced the names *Archocentrus, Parapetenia, Thorichthys*, and *Theraps* currently being suggested as possible subdivisions of the genus *Cichlasoma*.

That said, we have reflected the confusion that *does* exist within the *Cichlasoma* genus where it is relevant to fishkeepers and the identification and availability of aquarium stock. Such confusion often results from the supposed hybridization between species.

For each species featured in this section we provide details of its habitat, eventual size, diet, sex differences, and its compatibility and breeding potential in the aquarium. Such descriptions, together with the colour photographs that accompany them, serve to underline the exciting potential that the Central American cichlids represent in the fishkeeping hobby worldwide.

Cichla ocellaris

Eye-spot Cichlid
- **Habitat:** Rivers and lakes of Central and South America.
- **Length:** 600-700mm (24-28in).
- **Diet:** Juveniles (under 100mm/ 4in) are notoriously difficult to feed. Some exporters rely totally on live *Tubifex*. It may be possible to encourage small specimens to take earthworms and *Gammarus* shrimps. Large specimens (300mm/12in upwards) will normally starve unless offered live fish as prey, which may not appeal to many aquarists. Weaning large wild-caught *Cichla* away from taking live food is very difficult, and sometimes impossible.
- **Sex differences:** Juvenile fishes cannot be sexed using visual characteristics. Even large fishes do not show obvious differences, although males are thought to be more colourful.
- **Aquarium compatibility:** The Eye-spot Cichlid is a true fish predator and will undoubtedly consume any fish that will fit between its jaws. Despite this characteristic, *Cichla ocellaris* can be kept with large *Cichlasoma* species fairly easily. It does not appear to exhibit the same territorial aggression as *Cichlasoma* species, perhaps because of its pelagic, or mid-water, existence.

- **Aquarium breeding:** Commercial breeding may be practical, but few aquariums could accommodate a spawning pair. An unpublished report from West Germany suggests that a cichlid enthusiast has achieved spawning success, with the pair showing typical parental care.

A long, laterally compressed cichlid well suited to a predatory life. Its streamlined body shape enables it to put on impressive bursts of speed in pursuit of its fish prey.

Cichlasoma aureum

Gold Cichlid; Blue Red-top Cichlid
- **Habitat:** Southern Mexico and Guatemala.
- **Length:** 150mm (6in).
- **Diet:** *Cichlasoma aureum* is not difficult to accommodate because it will accept a wide range of prepared foods. It shows a preference for larval foods, such as bloodworm and gnat larvae. It will also take flake, foodsticks, frozen brine shrimps, *Mysis* shrimp and *Gammarus* shrimp.
- **Sex differences:** Males show dorsal and anal fin extensions,

Below: **Cichla ocellaris**
This juvenile specimen is the size and colour phase most likely to be encountered by fishkeepers. Small individuals are difficult to feed.

Above: **Cichla ocellaris**
*Adult specimens, such as this one,
will thrive in a large cichlid community.
This species prefers live food.*

although this is much easier to
see in mature pairs.
● **Aquarium compatibility:** This
species should be considered in
the same mould as the Firemouth
Cichlid (*Cichlasoma meeki*)
because its body shape and size
are almost identical. As a small to
medium-sized cichlid, it is well
suited to aquariums in the
90-120cm (36-48in) range.
● **Aquarium breeding:** A spawning
pair show an intensification of
colour, especially the male,
which can appear an iridescent
green. The pair clean and guard a
chosen spawning site – usually a

Below: **Cichlasoma aureum**
*A beautiful cichlid, but one that is not
widely available in fishkeeping circles.
This fine male specimen shows the
typical fin extensions.*

flat surface of slate or bogwood. Once the eggs are produced and fertilized, the parents fan the batch of up to several hundred with a sweeping movement of their ventral fins. The viable eggs hatch in 48-72 hours, depending on the aquarium temperature. The brooding female displays a lighter ventral area to enable the fry to identify her once they are free swimming. The parents dig several pits in the substrate and move the fry from site to site during the first week or so.

The red-tipped dorsal fin of *Cichlasoma aureum* gives this medium-sized cichlid an appealing colour pattern. Unfortunately, the Gold Cichlid (or Blue Red-top Cichlid as the author suggests its common name should be) is less widely available than some species.

Cichlasoma carpinte
Blue Texas Cichlid; Green Texas Cichlid
- **Habitat:** Mexico, River Panuco.
- **Length:** 300mm (12in).
- **Diet:** Shrimp, earthworms, insect larvae, foodsticks, pellets, flake food and, occasionally, leaf spinach.
- **Sex differences:** Males develop a head hump, which increases in

size as the fish reaches maturity. In breeding colours, males develop iridescent silvery blue markings on the body; females lighten in the ventral region.
- **Aquarium compatibility:** Dominant males can be destructive in a community aquarium, although they are ideal robust cichlids capable of holding their own among a group of large specimen fishes (i.e. one of each of a number of species).
- **Aquarium breeding:** Commercially bred specimens are frequently available, but few aquarists are interested in spawning this species. A mature pair will choose a spawning site and defend the area in typical cichlid fashion. Large broods are reported, with improved growth and survival rates resulting from removing some of the developing fry from the parents during the first two weeks of them becoming free swimming.

The dark blue of this species is instantly recognisable from the pale blue-green of *C.cyanoguttatum.*

Below: **Cichlasoma carpinte**
This species can be considered one of the most adaptable and robust of available Central American cichlids.

European fishkeepers know this form as the Texas Cichlid and the pale form *C.cyanoguttatum* as the Cuban Cichlid. The names are frequently reversed in the USA, where the Texan populations of *C.cyanoguttatum* (see page 46) are proudly known as America's National Cichlid.

Cichlasoma citrinellum
Red Devil; Midas Cichlid; Lemon Cichlid

- **Habitat:** Crater Lakes of Nicaragua, The Rio San Juan system.
- **Length:** 250mm (10in).
- **Diet:** Shrimp, earthworms, spinach, foodsticks and large flake food.
- **Sex differences:** Sexually mature males develop a distinctive hump on the head – a so-called 'nuchal' hump.
- **Aquarium compatibility:** Aggressive when adult.
- **Aquarium breeding:** Large specimens often clash violently during pre-spawning activity. Jaw locking and chasing are permanent behaviour patterns

Below: **Cichlasoma citrinellum**
This large male displays the head hump (or 'nuchal' hump) that is characteristic of adult Red Devils.

during this period. In a group of cichlids the pair will usually become territorial and vigorously defend a corner of the aquarium from all-comers. If they are sexually mature and ready to spawn, the two fishes will display breeding tubes – small points between the ventral and anal fins. Sometimes the pair will engage in extended cleaning of a rock or piece of bogwood. Alternatively, if the female is not ready, the male will attack her. This is shown by the female 'cowering' or being beaten into submission in the upper corner of the aquarium. In this situation, the female can suffer extensive body and fin damage. If the spawning is successful, several hundred eggs are produced. After 24-48 hours, the infertile eggs show signs of fungus attack; the parents will clean off the debris from unviable eggs, although sometimes the fungus spreads to developing eggs. Once the fry have emerged, the parents will often move them from site to site, picking each one off, one at a time. The fry are usually free swimming on the fifth or sixth day after spawning and will follow the parents around the spawning area as they search for food.

A problem exists relating to the identity of *Cichlasoma citrinellum* and *Cichlasoma labiatum*. The latter is thought of as a large-lipped species, usually displaying some orange-red pigment. *C.citrinellum* is found in yellow or red forms and is a simple-lipped species. The problem is caused by the polymorphism described clearly by Dr Paul Loiselle in commercial literature (following Professor Barlow's work published in 'Investigations of Ichthyofauna of the Nicaraguan Lakes') as the *Cichlasoma labiatum* species complex. Scientists have observed introgressive hybridization between *C.citrinellum* and *C.labiatum* in the smaller lakes, where the lack of suitable partners makes such hybridization the only viable alternative for survival.

Aquarium stock has developed from imports into the USA and Germany during the 1960s and 1970s. Therefore, the original genetic strength of colour, size and shape have become diluted. Nonetheless, *C.citrinellum* has made an indelible mark on the aquarium hobby.

Cichlasoma cyanoguttatum
True Texas cichlid
- **Habitat:** Mexico, in Rio Grande, Rio Pecos, Rio Conchos.
- **Length:** 250mm (10in).
- **Diet:** Shrimp, earthworms, insect larvae, foodsticks, leaf spinach and flake food.
- **Sex differences:** Males develop the frontal, or nuchal, hump and can be slightly stronger in colour.
- **Aquarium compatibility:** Large individuals can become dominant in a cichlid group and will relentlessly bully subdominant or weaker fish. Spawning pairs undoubtedly require isolation from a community aquarium.
- **Aquarium breeding:** Spawning pairs are dark in the mid- body/caudal region, light from the head back to the centre of the body. The female prepares the spawning site while the male chases off intruders. The parents

Above:
Cichlasoma cyanoguttatum
The True Texas Cichlid is widely available and an ideal candidate for large cichlid community systems.

share responsibility for cleaning the eggs, although the female takes on the major part of the work. The female digs pits in which to place hatching fry and frequently moves them from site to site. Fry growth is dependent on the frequency and amount of feeding. Rearing a proportion of the fry separately ensures that a reasonable number develop.

This species was confused in Europe under a false name, '*C.tetracanthus*'. Its name has been long associated with the Blue Texas Cichlid (*C.carpinte*) and will continue to be so for many years because of the name confusion in commercial literature.

Cichlasoma dovii
Dow Cichlid; Dow's Cichlid; Wolf Cichlid
- **Habitat:** Lake Nicaragua, Costa Rica, Honduras. In clear water areas near submerged or waterlogged brush or tree trunks.

- **Length:** 500-700mm (20-28in).
- **Diet:** Aquarium-raised specimens will accept a wide range of shrimp, prawn and earthworm-sized foods. Large specimens will eat fishes, so it is important to keep them with large cichlids.
- **Sex differences:** Males often display a more intense colour pattern and a larger, more pointed, anal fin, although this can be considered only as a guideline in most cases.

- **Aquarium compatibility:** The Dow Cichlid's attainable size puts this fish into the *superclass*, and excludes it from all but the largest aquarium. Few aquarists have kept this cichlid with other community cichlids; its comparative rarity solves the problem for most fishkeepers.
- **Aquarium breeding:** Aquarium spawnings are rare but not unheard of among cichlid enthusiasts. Large broods are reported, with typical parental care being observed.

Cichlasoma dovii has been sporadically available over the years and is well known among American Cichlid Association members through a large specimen named 'Pablo' kept by an enthusiast. It has been reported in popular USA magazines by Dr Paul Loiselle that the fish was as well known to the ACA membership as its owner. This shows that large individual cichlids have character and can become true pets.

It is interesting to note that the Dow Cichlid is a major predator of *Neetroplus* in Nicaraguan Lakes.

Some authors recommend a 200-1 ratio of fish to tank space to maintain an adult Dow Cichlid.

Below: **Cichlasoma dovii**
A pair of Dow Cichlids in typical spawning colours. The male (above) is clearly larger than the female.

Cichlasoma friedrichsthalii

Friedrichsthal's Cichlid

- **Habitat:** Widespread throughout Mexico, Guatemala, Honduras, Belize, Costa Rica and Nicaragua.
- **Length:** 250mm (10in).
- **Diet:** Although a fish eater in its natural habitat, young farm- or tank-raised specimens will take almost any prepared food.
- **Sex differences:** Males are larger than females, display a more ornate body speckling and sport elongated dorsal and anal fin rays.
- **Aquarium compatibility:** As with all of the large *Cichlasoma* species, this fish is best kept with similarly sized cichlids, such as *C.synspilum* and *C.maculicauda*. Juveniles are similar in appearance to *C.motaguense* and are also reminiscent of *Petenia* youngsters.
- **Aquarium breeding:** Aquarium spawnings are rarely mentioned in literature, although it is reasonable to assume that this species spawns in much the same way as closely related forms, such as *C. motaguense*.

Above: **Cichlasoma hartwegi**
An adult spawning pair of Tail Bar Cichlids with fry. The male is the upper and noticeably larger fish of the pair. This species, although a relative newcomer to fishkeeping, has proved fairly easy to spawn, which should ensure its continuing availability around the world.

Below: **Cichlasoma friedrichsthalii**
A female specimen displaying the characteristic lines and blotches extending from the eye to the tail.

Cichlasoma friedrichsthalii is frequently confused with *Cichlasoma motaguense,* from which it can be distinguished by the vertical bars on either flank. A series of black blotches blend into this pattern between the base of the caudal fin and the eye. This part of the pattern is shared with *C.motaguense.*

Cichlasoma hartwegi

Tail Bar Cichlid

- **Habitat:** Mexico, in the Rio Grijalva over sand and rock substrate.
- **Length:** 150mm (6in).
- **Diet:** Shrimp, insect larvae, leaf spinach and flake food.
- **Sex differences:** Males in spawning/brood-caring colour are silver with a hint of red speckling; the posterior half of the body has black criss cross markings. The caudal base has a characteristic broad stripe (sometimes extending in an inverted arc from tail to eye), which the author refers to in the suggested common name.

- **Aquarium compatibility:** This is not known to be an aggressive species and would seem ideally suited to small to medium-sized community aquariums.
- **Aquarium breeding:** Successful spawnings have been recorded. It is said to be an easy species to breed, with typical parental care being shown by the breeding pairs. Moderately sized broods are produced; you can expect about 100-200 fry.

This species is similar to *Cichlasoma fenestratum*, which also comes from Mexico and grows to 200mm (8in), with pink to red fin edges and strong pink markings on the head. *C.fenestratum* and *C.hartwegi* have vertical body stripes, seemingly more pronounced in the former species, although the caudal peduncle bar is distinctive in *C.hartwegi*. The Tail Bar Cichlid was described to science in 1980, and so it is a relative newcomer to the list of Central American cichlids entering the fishkeeping hobby.

49

Above: **Cichlasoma intermedium**
A peaceful cichlid that has been made available through the efforts of German importers and fishbreeders.

Above right: **Cichlasoma labiatum**
A close-up showing the pronounced lips of an adult male – a distinctive characteristic of the species.

Cichlasoma intermedium
Intermedium Cichlid
- **Habitat:** Rivers in Mexico, Belize and Guatemala.
- **Length:** 150mm (6in).
- **Diet:** Shrimps, earthworms, foodsticks and flake food.
- **Sex differences:** Difficult to establish, although sexually mature males are more slender and can display fin extensions.
- **Aquarium compatibility:** An excellent small to medium-sized species ideal for a Central American cichlid community aquarium.
- **Aquarium breeding:** Aquarium-raised specimens are not commonplace, although a reported spawning has resulted in this species becoming available in limited quantities. The spawning pattern is typical of closely related smaller species of *Cichlasoma*. The fry are said to develop well from the free-swimming stage on newly hatched brine shrimp; most fish fry will accept this live food.

This is a relatively new species in aquarium circles. The author encountered a specimen of

C.intermedium for the first time in 1984, when aquarium-raised specimens became available from West Germany. The basic brown-green colour sometimes glistens as a border to the unique, reversed L-shaped body patterning.

Cichlasoma labiatum
Large-lipped Cichlid; Red Devil
- **Habitat:** Lakes in Nicaragua.
- **Length:** 200-250mm (8-10in).
- **Diet:** A great deal of research has been made into the feeding patterns of members of the *C.labiatum* group. In lakes and rivers, they are recorded as feeding on snails, organic debris, small fishes, aquatic insects and fish eggs. In aquariums, this can be mirrored with frozen or freeze-dried foods.
- **Sex differences:** Males display pronounced lips and often have long, extended anal and dorsal fin rays.
- **Aquarium compatibility:** Apart from a continual desire to dig into the aquarium substrate, the Red Devil is as compatible within a Central American cichlid community as any comparably sized species.

● **Aquarium breeding:** Aquarium-raised specimens spawn easily, although parental care appears to diminish rapidly within the first week of the fry becoming free swimming. Removing the fry into a rearing tank is essential. Spawning males can prove to be extremely aggressive and excessive jaw locking, tail beating and biting can leave a female very much the worse for wear. Large fishes can produce up to 7500 eggs!

Several colour forms exist, including white, yellow, yellow-orange and red. The large-lipped form with a red body is said to be extremely scarce in nature and correspondingly rare in aquarium circles. Juveniles are grey to green, with several cross bands. Hybridization has occurred between this species and *C. citrinellum.* (See page 45.)

Below: **Cichlasoma labiatum**
This colour form is the one most likely to be seen by fishkeepers. Red-bodied specimens are rare.

Cichlasoma longimanus
Rose-breasted Cichlid

- **Habitat:** Mexico (Rio San Juan Basin), Guatemala, Honduras and Costa Rica, especially warm mud-bottomed backwaters.
- **Length:** 200mm (8in).
- **Diet:** Crustaceans, insect larvae and prepared flake foods.
- **Sex differences:** Females have a distinctive colour pattern in the dorsal region extending into the dorsal fin. (This characteristic has even led to the sexes being confused as different species by some hobbyists.) Adult males display distinctive lateral line blotches and possess longer dorsal and anal fins than the female fishes.

- **Aquarium compatibility:** This is not an aggressive species, although a breeding pair will spawn in the open on the substrate and, to quote Paul Loiselle retelling a friend's experience with this fish, they will plonk the eggs down on a solid surface and dare any aquarium fish present to do something about it!
- **Aquarium breeding:** As a substrate sifter, the Rose-breasted Cichlid will choose a site on the gravel bed and dig out a pit. Up to 500 eggs are deposited, more or less in a pile. In the wild, they are said to choose sites within rocky areas. Males will spawn with several

females, joining in brood protection until the fry are free swimming, before entertaining the next female.

American stock originated from Lake Jiloa in Nicaragua, and Costa Rica, but the species has yet to become established in the hobby.

Cichlasoma maculicauda

Black Belt Cichlid

● **Habitat:** Southern Mexico, Guatemala, Belize, Costa Rica and Panama.
● **Length:** 300mm (12in).
● **Diet:** Only tank-raised fishes are available and these will feed on almost any aquarium prepared food. Mature fishes will relish *Gammarus* shrimp and large earthworms.
● **Sex differences:** Sexing juvenile Black Belt Cichlids can be achieved only with a certain amount of educated guesswork. In groups of semi-adult specimens (about 150mm/6in), the males exhibit early sexual behaviour as they establish a pecking order and display to the females. Adult males are slender in direct comparison to females, often displaying longer anal and

Below: **Cichlasoma longimanus**
A pair of Rose-breasted Cichlids, (the male is the upper fish). An attractive newcomer to the hobby.

dorsal fin rays and showing stronger red in the caudal fin.
- **Aquarium compatibility:** This is a typical *Cichlasoma* species of medium to large size, once much sought after by aquarists. It is ideal for a general community system of cichlids. Its colour pattern of red cheek and caudal fin will enhance any Central American cichlid display.
- **Aquarium breeding:** Aquarium spawnings have been achieved, although not frequently, as in the case of *C.nicaraguense* and *C.synspilum*. Spawning males darken under the head and forward ventral area whereas females display a white ventral region after spawning. The distinctive colour change in the female helps to attract fry so that the brood can be controlled.

Juveniles rarely show the red patterning, although the distinctive vertical mid-body stripe and lateral caudal peduncle bar make them

Below: **Cichlasoma maculicauda**
This male specimen displays the beautiful red patterning that ensures the popularity of this species among cichlid enthusiasts worldwide.

quite distinguishable from similar species, such as *C.synspilum*, which lacks the mid-body vertical bar pattern of *C. maculicauda.*

Cichlasoma managuense
Jaguar Cichlid; Managua Cichlid
- **Habitat:** Honduras, Nicaragua and Costa Rica. In turbid slow-moving waters, especially in small tributaries of large rivers. Found close to rock substrate in Lake Nicaragua.
- **Length:** 300mm (12in).
- **Diet:** As the fishes available are usually small tank-raised specimens, they will eat virtually any prepared foods, although it is not advisable to feed them totally on a diet of pellets or foodsticks as they will tend to produce a considerable amount of organic debris in the system. A diet of fish, shrimp and earthworms will ensure fast growth.
- **Sex differences:** Adult males appear more ornately patterned than females and usually display extended anal and dorsal fin rays.
- **Aquarium compatibility:** Large specimens can be particularly aggressive among their own kind and to smaller fishes, but

Above: **Cichlasoma managuense**
An adult male Jaguar Cichlid in dominant colour form – the highlight of any large cichlid community.

youngsters are reasonably compatible with similarly sized fishes in the aquarium.
● **Aquarium breeding:** Successful aquarium spawnings are recorded, usually at the expense of other aquarium occupants, which are relentlessly harassed by the parent fishes in protection of their offspring.

The aquatic hobby confused this species with *Cichlasoma motaguense* for many years. However, the silver background colour in the Managua Cichlid (particularly bright in the colour form known as the Jaguar Cichlid) is not seen in the Motaguense Cichlid, which has a yellow-brown base to its overall colour patterning.

Cichlasoma meeki
Firemouth Cichlid
● **Habitat:** Rivers in Mexico (Yucatan) and Guatemala.
● **Length:** 150mm (6in).
● **Diet:** Bloodworms and shrimps are favourite foods for Firemouths, although they will accept a wide range of prepared foods in the aquarium.
● **Sex differences:** Males are more colourful than females, have larger extensions to the dorsal

and anal fins, and are often seen blowing out their throat and gill membranes (branchiostegals) in breeding displays.
● **Aquarium compatibility:** The Firemouth Cichlid is one of the first species from the Central American group to be encountered by fishkeepers. To aquarists used to the aggressive behaviour of larger *Cichlasoma*, this small species would appear a peaceful fish. In a general community aquarium, however, aquarists would consider the Firemouth Cichlid a bully; it is capable of consuming smaller tropical fishes. It is suited to the smaller aquarium and, when adult, can be kept with larger Central American cichlids.
● **Aquarium breeding:** A spawning pair show an intensification of colour, with the male developing a particularly bright red throat and belly – hence the common name. Once an area has been selected, the male defends the territory, chasing off all other fishes while he impresses the female with his 'peacock' style of display, which involves blowing out his gills and displaying his fins. The actual spawning site (usually a rock surface) is cleaned and protected by the pair, and then up to 500 eggs are produced by the female and fertilized at once by the male. Once hatched, the fry are kept in hollows dug out

by the pair until they are free swimming and capable of following the adults. The fry can be fed freshly hatched brine shrimp and powdered flake food. This, together with the cloud of chewed food the parents spit out after feeding, should see them safely through the 14-21 day stage, when they should be removed to a separate aquarium for growing on.

Firemouth Cichlids were first imported into the USA in about 1915 and into Europe in the 1930s. The present generation of aquarium fishes are commercially bred in the Far East and the colour exaggerated by hormone feeding.

Cichlasoma melanurum
Black-blotch Cichlid
- **Habitat:** Lakes in Guatemala; rivers in Belize.
- **Length:** 200mm (8in).
- **Diet:** Crustaceans, insect larvae, leaf spinach and flake food.
- **Sex differences:** Overall, females are less pigmented and have shorter fin rays than males. Adult pairs show a darkening of the ventral region.

Above: **Cichlasoma meeki**
A male Firemouth Cichlid, guaranteed to add striking colour to a community of small to medium-sized cichlids.

- **Aquarium compatibility:** This species is comparable to *C.maculicauda* and *C.synspilum* in its general temperament, and would be suited to community systems containing these species.
- **Aquarium breeding:** Very little has been published on the spawning of the Black-blotch Cichlid; they are reported to dig out gravel from under bogwood and spawn on the wood face. Typical parental care is shown by this fish; brood caring females are said to show intense colouring.

As a newcomer to the aquarium hobby, this species is not widely available. Its colour pattern is almost intermediate between *C.maculicauda* and *C.synspilum*.

Cichlasoma motaguense
Motagua Cichlid
- **Habitat:** Mexico, southern Guatemala (Rio Motagua), and El Salvador. A noted river species.

Above: **Cichlasoma melanurum**
Similar in general appearance to the better-known Cichlasoma synspilum, *this medium-sized species has gained popularity over recent years.*

- **Length:** 300mm (12in).
- **Diet:** Very much the typical large cichlid diet of whole shrimp, large chopped earthworms, foodsticks and pellet food.
- **Sex differences:** Adult males are more ornate in colour pattern than females and sometimes exhibit slightly enlarged head fronts. Sexually mature females in breeding colours display indistinct vertical bands.
- **Aquarium compatibility:** This *Cichlasoma* is certainly a boisterous cichlid when over half grown, but it is possible to keep them successfully in a large community of similarly sized cichlids.
- **Aquarium breeding:** Several spawnings have been recorded, and large broods have been successfully raised.

Cichlasoma motaguense has been confused with *C.managuense* by importers and for a period the two names have been interchanged. The two fishes have a similar lateral blotch pattern, but *C.motaguense* has a yellowish body hue and retains the blotch line in the adult. By contrast, *C.managuense* displays an iridescent, almost reticulated, silver-black pattern.

The close similarity in patterns between *C.motaguense* and *C.friedrichsthalii* has also often led them to be confused in literature. A breeding female *C.motaguense* even shows the vertical bands distinctive in *C.friedrichsthalii*. Male gold forms are recorded in literature and, according to Dr Paul Loiselle, are known as 'El Rey de la Guapotes' – King of the Guapotes – in Nicaragua.

Below: **Cichlasoma motaguense**
A young specimen of this widely available species, showing the colour pattern and overall body shape most likely to be seen by fishkeepers.

Cichlasoma nicaraguense

Nicaragua Cichlid

- **Habitat:** Nicaragua (Lake Managua) and Costa Rica.
- **Length:** Males 250mm(10in); females 200mm(8in).
- **Diet:** Juveniles and adults accept a wide variety of prepared and frozen food, and particularly relish chopped earthworms and fresh leaf spinach.
- **Sex differences:** This species is fairly easily sexed in maturity; the females are slightly smaller than males and retain a simpler nonetheless bright coloration of yellow and blue-green. A lateral line black body stripe – prominent in juveniles – is retained by the female, but is seen as an indistinct central body spot in the predominantly yellow adult male.
- **Aquarium compatibility:** This medium to large species is ideal for a large cichlid community aquarium. It is not one of the most aggressive species of Central American *Cichlasoma* available, but is still capable of causing considerable damage in a small fish community. Ideally, keep this species with *C.synspilum* and *C.maculicauda* until they sex out in size and colour, when surplus fishes may be removed if necessary.
- **Aquarium breeding:** The Nicaragua Cichlid has proved easy to spawn once a compatible pair have formed a lasting bond. A deep pit is dug in the substrate and the eggs placed into it in a clump. Fishkeepers have found it best to leave the parents to raise the fry, as this encourages future breeding success. In one instance, an aquarist removed the eggs to raise them artificially and noticed that the pair became aggressive towards one another.

  Egg numbers depend on the maturity of the parents, but reports suggest a spawn of 300-500 eggs is not uncommon, with likely hatching rates between 20 and 50 percent. Immature males will cause a greater number of

Above: **Cichlasoma nicaraguense**
A pair (the male above). These fishes are extremely popular among fishkeepers for their brilliant colours.

eggs to be infertile. Therefore, early spawnings can be expected to produce lower hatching rates. The eggs of this species are reported to be non-adhesive, a characteristic unique among Central American cichlids.

This brightly coloured cichlid took European cichlid enthusiasts by storm when it first appeared in 1979/80. The colours on sexually mature pairs – brilliant greens, yellows and blues – rival those of many coral reef fishes. The female displays a blaze of colour normally associated with outstanding males in some cichlid species. The superb adult colour pattern of the Nicaragua Cichlid has undoubtedly

caused a great upsurge of interest in keeping Central American cichlids. Juvenile fishes are plain silver with a black lateral stripe and mid-body blotch, and rarely herald the outstanding beauty of the adult.

Cichlasoma nigrofasciatum
Convict Cichlid; Zebra Cichlid
- **Habitat:** Guatemala, El Salvador, Honduras, Nicaragua and Costa Rica.
- **Length:** 100mm (4in).
- **Diet:** The Convict must be the easiest species to accommodate on the dietary front; juveniles and adults will accept every form of tropical fish food. By enhancing the diet with bloodworm, chopped earthworms and frozen shrimp, a pair can be brought easily into spawning condition.
- **Sex differences:** Juvenile Convict Cichlids are difficult to separate into males and females, although the former tend to be more aggressively active when approaching sexual maturity. Also, adult males display particularly intense black vertical bands. These are less prominent in females, which – especially in breeding dress – are orange-yellow in the ventral region. And males invariably possess larger dorsal and anal fins that usually extend into filaments.
- **Aquarium compatibility:** Few community aquariums have not been plunged into chaos by the introduction of a renegade cichlid by the novice fishkeeper. Most aquarists will encounter cichlids for the first time in this manner and the most likely species to be purchased in ignorance of its aggressive tendencies is undoubtedly the Convict Cichlid. The aggression stems from the

cichlid's desire to spawn and protect its progeny. While it may infuriate the irate fishkeeper, who hates to see community fishes victims of serious assault, this behaviour ensures that the fish will be successful in raising its fry, both in the natural habitat and in the aquarium.

● **Aquarium breeding:** This is one of the easiest species of *Cichlasoma* to spawn and raise. A breeding pair will choose a vertical or horizontal spawning site and defend the area vigorously from all-comers, including cichlids larger than themselves. Commercially bred generations are said to be less protective of free-swimming fry and lose interest in the brood within a week, in some instances. It is wise to remove at least 50 percent of the free-swimming fry to a separate raising system.

The zebra-like banding pattern is highlighted on the spawning male whereas the female develops a yellow underside to attract the brood.

Below: **Cichlasoma nigrofasciatum 'Kongo'**
The gold form of this species has become well established in the USA.

Wild-caught specimens, said to be highly coloured (especially in spawning condition), are rarely encountered. The Convict Cichlid is commercially bred extensively in the Far East and also in the USA, where a gold variety is well established in the hobby.

Cichlasoma octofasciatum
Jack Dempsey Cichlid
● **Habitat:** Mexico (Yucatan), Guatemala and Honduras.
● **Length:** 200mm (8in).
● **Diet:** Commercial or tank-bred specimens will feed on a wide range of foods, although they show a preference for crustaceans, shrimps, snails, etc. Large specimens will greedily take foodsticks, whole shrimps and earthworms. This species will also show interest in green foods, such as leaf spinach or lettuce, but this should be offered only occasionally.
● **Sex differences:** Sexually mature males are blue-black in colour; females tend to be paler in their markings.
● **Aquarium compatibility:** Of the small to medium-sized cichlids, the Jack Dempsey Cichlid is probably the most pugnacious,

Above:
Cichlasoma nigrofasciatum
An adult male, showing the normal colour form of the Convict Cichlid.

Below: **Cichlasoma octofasciatum**
This ebullient cichlid was one of the earliest species made available to fishkeepers. Striking blue spangling.

hence its common name. It is entirely suited to smaller cichlid communities if kept with Firemouth, Convict and Salvin's Cichlids. (See pages 55, 59, 65.)
● **Aquarium breeding:** Aquarium spawnings are said to have produced broods of up to 800 fry, although aquarium-raised breeding pairs are more likely to produce and raise much smaller broods. As the fry begin to show dark stripes on the body, remove them to a separate raising tank before the parent fishes begin the spawning cycle again and lose interest in protecting the brood.

Known under the name of *Cichlasoma biocellatum* for some time in commercial literature, the Jack Dempsey is a popular species among newcomers to cichlids.

61

Cichlasoma panamense
Panama Cichlid
- **Habitat:** Small tributaries of the Chagres and Bayano Rivers in Panama.
- **Length:** 100mm (4in).
- **Diet:** A typical cichlid diet should enable the fishkeeper to develop the aquarium-raised specimens of this species that are available from time to time.
- **Sex differences:** The adult male is larger than the female and, especially in breeding dress, displays a red body hue broken by a dark line of vertical blotches from the caudal base to the mid-point of the body. Brood-caring females display a pale ventral region and a broken vertical line of blotches on the body.
- **Aquarium compatibility:** Ideally suited to the smaller cichlid community system, this tenacious yet unaggressive species is one of the most recent newcomers to the European aquarist market.
- **Aquarium breeding:** Aquarium spawnings have been widely reported. Spawning pairs are said to prefer caves and plant pots as breeding sites. Brooding females display a light and dark body pattern, presumably for fry recognition and protection. This species was originally allied with

Above: **Cichlasoma panamense**
A juvenile, displaying the plain body colour reminiscent of Neetroplus nematopus. *Peaceful and easily bred.*

Neetroplus and, from aquarium keeping aspects, can be considered similar to some extent. It appears that the Panama Cichlid does not share the awesome aggression of *Neetroplus* but is capable of producing and raising similarly sized broods in the aquarium.

Juveniles have a uniform grey body, which does little to recommend the species initially, but its small size and relatively peaceful nature will ensure its popularity as an ideal small community aquarium cichlid.

This species is a relative newcomer in comparison to other well-known forms; it became widely available for aquarium use in 1983 following a scientific survey carried out in the rivers of Panama.

Cichlasoma robertsoni
Metallic Green Cichlid
- **Habitat:** Lakes, lagoons and rivers in Mexico, Guatemala and Honduras.
- **Length:** 200mm (8in).
- **Diet:** Crustaceans, insect larvae and flake food.

Above: **Cichlasoma panamense**
An adult, illustrating the stark contrast in patterning when compared to the somewhat plain juvenile.

● **Sex differences:** Males are larger and display a brighter coloration than females.
● **Aquarium compatibility:** A moderate cichlid by normal *Cichlasoma* standards, this species is suitable for small to medium-sized aquarium communities.
● **Aquarium breeding:** Habitat observations reveal that this cichlid holds territories close to submerged tree trunks and large rocks in river conditions. It is

capable of breeding in a cichlid community, even one containing *Neetroplus nematopus*, according to Dr. Paul Loiselle.

The Metallic Green Cichlid is similar in shape and general appearance to the Firemouth Cichlid (*Cichlasoma meeki*), but differs in colour pattern and body spot. It has metallic green scales that catch the light, making it a most beautiful newcomer to the fishkeeping hobby. It is a substrate sifter and tends to dig continually into the aquarium gravel.

Below: **Cichlasoma robertsoni**
A handsome new cichlid, but not yet widely available to fishkeepers.

Cichlasoma sajica
Sajie's Cichlid
- **Habitat:** Small streams and lakes in Costa Rica.
- **Length:** 125mm (5in).
- **Diet:** A wide variety of foods will keep this small cichlid in good condition in the aquarium.
- **Sex differences:** Adult males possess extended dorsal and anal fins and show a blue tinge to the body, whereas females are smaller and yellowish. Brooding females darken and display a thin vertical body stripe from the mid-dorsal region.
- **Aquarium compatibility:** It is ideal for a 60-90cm (24-36in) cichlid community aquarium, although sadly it is less widely available than the cichlid most confused with it, *C.spilurum*. Sexually active pairs can be extremely unsociable towards other aquarium occupants and should be removed to a special breeding tank. If kept alongside larger cichlids, they will act as small food scavengers and will be rarely threatened by their larger cousins in the system.
- **Aquarium breeding:** This species is known as an open spawner, although observations show that it will take cover in pots or rock caves.

Cichlasoma sajica could easily be confused with *C.spilurum* and *C.spinosissimum*, although the red

Below: **Cichlasoma sajica**
A male specimen showing the clear similarity in shape of this rarely encountered cichlid to the widely obtainable Cichlasoma spilurum.

speckling in the finnage of sexually mature males would distinguish it fairly easily. (Also, the central vertical bar is usually prominent in juveniles and adults and is a characteristic of the species.) They are also fairly distinct in their geographical distribution: *C.sajica* is found in Costa Rica; *C.spilurum* in Guatemala and Belize; and *C.spinosissimum* in the highland rivers of Guatemala.

Cichlasoma salvini
Salvin's Cichlid; Tricolor Cichlid
- ● **Habitat:** Mexico, Guatemala, Honduras.
- ● **Length:** 150mm (6in).
- ● **Diet:** This cichlid shows a preference for larval or shrimp foods and will retain its good colour if fed regularly on bloodworm and *Gammarus* shrimp.
- ● **Sex differences:** Sexually mature males only two-thirds grown develop a red blotch in the ventral body/anal fin region, which makes this one of the most attractive of the smaller cichlids.
- ● **Aquarium compatibility:** A tenacious dwarf Central American cichlid, Salvin's Cichlid will enhance any community, large or small. In certain

Above: **Cichlasoma salvini**
A striking female specimen that shows clearly why the beautiful Tricolor Cichlid is one of the most popular species available.

community aquariums, however, this beautiful cichlid would be bullied because of its distinctive pattern and smaller size. Breeding pairs can be very aggressive and will bite other aquarium cichlids viciously unless given an aquarium of their own for breeding activites.
- ● **Aquarium breeding:** This species is said to prefer placing eggs on a sloping or vertical surface. Egg numbers have been recorded at 1000, although half that figure is the accepted norm. Both parents show bright colour patterns, especially while they are protecting the eggs and fry. Other fishes are said to associate the bright colour with aggression and endeavour to avoid them during this period. Fry hatch within three days and will soon take newly hatched brine shrimp and powdered food. Experience shows that it is best to remove fry from the parents and separate the parent fishes to save the female from the somewhat over-

enthusiastic attentions of the male in his desire to repeat the spawning cycle *before* the female is ready to start breeding again.

The Tricolor Cichlid is not widely known in fishkeeping circles, but it is an ideal species to consider for a breeding programme. It is well suited to small to medium-sized aquariums because it reaches sexual maturity at the relatively small size of 100mm(4in).

Cichlasoma sieboldii
Bandit Cichlid
- **Habitat:** Pacific slopes of Costa Rica and western Panama. It lives in fast-flowing or still waters.
- **Length:** 100-150mm (4-6in) in aquarium specimens, although the maximum length recorded in literature is 300mm (12in).
- **Diet:** Crustaceans, insect larvae, vegetables, lettuce, leaf spinach and flake food. In the wild, it feeds by combing insect larvae from algae on rocks.
- **Sex differences:** Females are noticeably smaller than males.
- **Aquarium compatibility:** This can be an aggressive species,

which suggests that it may be best kept with medium-sized *Cichlasoma* species.
- **Aquarium breeding:** Breeding pairs develop brooding colours that give the fish its common name. Lines across the forehead extend around the eyes to give the appearance of a fish wearing a mask. The throat turns dark grey, this colour being carried into the body through about seven coalescing blotches that extend from the pectoral fins to the tail. *C.sieboldii* is more or less an open spawner, although it will take advantage of bogwood or rock areas on which to lay eggs.

Cichlasoma sieboldii undergoes almost a 'Jeckyll and Hyde' transformation when breeding by changing both its body colour – normally sandy overall with small red spots – and personality.

Cichlasoma spilurum
Blue-eyed Cichlid; Jade-eyed Cichlid
- **Habitat:** Guatemala and Belize.
- **Length:** 125mm (5in).
- **Diet:** This species will thrive on

Above: **Cichlasoma spilurum**
A male specimen of the Jade-eye Cichlid. This is an excellent species to

much the same fare as other small to medium-sized cichlids, and will particularly relish bloodworm and *Daphnia*.

Below: **Cichlasoma sieboldii**
The Bandit Cichlid has recently become available to fishkeepers and is ideal for small aquariums.

introduce fishkeepers into the fascinating world of Central American cichlids. An easy species to breed.

● **Sex differences:** Males display a yellowish ventral region and are generally longer in the body and in the dorsal and anal fins than females. Brood-caring females have a distinctive series of vertical black bars, not dissimilar to those of the Convict Cichlid, (*C. nigrofasciatum*).
● **Aquarium compatibility:** Spawning couples – in keeping with the tenacious smaller cichlids – can be difficult; otherwise they are perfectly suited to small systems.
● **Aquarium breeding:** Breeding Jade-eyed Cichlids is relatively easy. Mature pairs will accept an upturned plant pot or rock cluster as a spawning site. Large broods can be expected and free-swimming fry should be removed to separate quarters when they are 14 days old.

The Jade-eyed Cichlid is probably the most underrated of the small Central American species. A spawning pair display warm colours and contrasting patterns that merit greater interest being shown in this species. They prove excellent parents and will successfully raise broods within a community aquarium without too much disruptive behaviour.

67

Cichlasoma spinosissimum

False Blue-eyed Cichlid; False Jade-eyed Cichlid

- **Habitat:** Rio Polochic in the highlands of Guatemala.
- **Length:** 125mm (5in).
- **Diet:** Crustaceans, insect larvae and most types of prepared foods.
- **Sex differences:** Adult males display a blue iridescence and have longer anal and dorsal fin rays than females.
- **Aquarium compatibility:** This species is ideally suited to small cichlid community systems.
- **Aquarium breeding:** Care-brooding pairs display vertical bands or stripes similar to those of the Convict Cichlid, (*Cichlasoma nigrofasciatum*), and the female darkens considerably in the ventral region. The spawning sequence is identical to that of *C.spilurum*, the cichlid it closely resembles.

This species is rather obscure in aquarium circles and probably confused in commercial literature. It may become available for the general aquarium market through the efforts of American or German collectors.

Above:
Cichlasoma spinosissimum
This species is thought to be a colour morph of the similar C. spilurum.

Below:
Cichlasoma spinosissimum
When available, treat it in the same manner as Cichlasoma spilurum.

Cichlasoma synspilum

Firehead Cichlid; Quetzal
- **Habitat:** Guatemala and Belize.
- **Length:** 300mm (12in).
- **Diet:** Young specimens will take almost any prepared food. Adults can become fussy, preferring large prawns, pellets, leaf spinach and earthworms.
- **Sex differences:** Adult males are brighter in colour than females and develop a slight nuchal hump. Juveniles can be sexed at the half-grown stage using the developing males' brighter colour pattern and fin development as a guide.
- **Aquarium compatibility:** This brightly coloured species will enhance any large Central American cichlid community. Together with *C.nicaraguense*, *C. synspilum* must take a good deal of the credit for promoting Central American cichlids to the aquarium hobby. If kept with similarly sized species, Firehead Cichlids will co-exist peacefully, although once a sexually mature pair have formed a bond, greater aggressive behaviour can be expected.
- **Aquarium breeding:** A spawning pair can produce large broods, although the difficulty lies in finding two compatible specimens. Pair bonding can result in continual spawning cycles. Fry rearing is best undertaken by removing 50-75 percent of the brood to a

Above: **Cichlasoma synspilum**
A female specimen, illustrating how vibrant this widely available species can be. Males display nuchal humps and an even brighter colour.

separate aquarium. If some fry are allowed to progress with the parents, there will be a high mortality rate in the period between 7 and 14 days after hatching, but continued parental care and pair bonding will be ensured by taking such action.

Cichlasoma synspilum is one of the most widespread of the commercially known larger species and is the species most likely be available at your local dealer. To develop a group of *C.synspilum* with the priority of creating a good spawning pair, start by buying between four and six juveniles. If possible, buy half of these from a different source to prevent close inbreeding. Raise the group in a 120×45×30cm (48×18×12in) aquarium or similarly sized system. As the cichlids become sexually active, the dominant male will display to the best female. At this stage, take the balance of the subdominant fishes back to the retailer as surplus to requirements. Although you will lose a good percentage of the original purchase price, you will have secured a compatible pair that should produce reasonably sized broods for you to raise and sell to offset the loss.

Cichlasoma trimaculatum

Shoulder-spot Cichlid; Three-spot Cichlid

- **Habitat:** Lagoons, lakes and rivers in Mexico, Guatemala and El Salvador.
- **Length:** 300-350mm (12-14in).
- **Diet:** Large shrimp, earthworms, foodsticks and pellet food.
- **Sex differences:** Males have extended finnage and display a stronger spot pattern than females. Males are larger and deeper in the body and generally

Above: **Cichlasoma trimaculatum**
A wild-caught male displaying the distinctive colour patterns reflected in its common name.

more colourful than females, with a claret-coloured throat.

- **Aquarium compatibility:** As a large species, the Shoulder-spot Cichlid should be considered

Below: **Cichlasoma trimaculatum**
A spawning pair in typical breeding coloration. The male is the upper fish. Aggressive when breeding.

suitable only for large aquariums containing similarly sized cichlids.

- **Aquarium breeding:** As an open spawner, the parent fishes select and defend their spawning site with a great deal of aggression.

This species has become widely available in recent years and earns its common name from the distinctive pattern of three body blotches, one of which is situated in the region of the shoulder.

Cichlasoma umbriferum
Blue Speckled Cichlid
- **Habitat:** Rivers in Panama and Colombia.
- **Length:** 300-400mm (12-16in).
- **Diet:** Crustaceans, fish, large earthworms, pellets and foodsticks.
- **Sex differences:** Females are clearly smaller and less brightly coloured than males.
- **Aquarium compatibility:** This species can be considered as safe only if it is kept with cichlids of its own size.

Below: **Cichlasoma umbriferum**
A sexually mature male of good colour. Provide plenty of swimming space in the aquarium.

- **Aquarium breeding:** This species is said to be an open spawner, although few reports exist of aquarium spawnings.

It is known that the Blue Speckled Cichlid is an open-water swimmer and thus requires a spacious aquarium. The bright blue spots on the greyish yellow body are distinctive to the species. Two black spots – one central on the body, the other on the tail base – help to identify juvenile fishes.

Cichlasoma urophthalmus
Mexican Tail-spot Cichlid
- **Habitat:** Lagoons in Mexico (Yucatan), Guatemala, Belize and Honduras.
- **Length:** 200mm (8in).
- **Diet:** Shrimp, earthworms, foodsticks and pellets.
- **Sex differences:** Males display extended fins and grow larger than females.
- **Aquarium compatibility:** Contrary to some descriptions, the author has found this species to be extremely aggressive. An adult specimen chased and attacked a larger Oscar (*Astronotus ocellatus*), until the latter had to be removed. This

particular incident occurred in a 240×60×60cm (96×24×24in) aquarium containing a selection of cichlids.

● **Aquarium breeding:** Breeding pairs prefer to spawn under cover of bogwood or rockwork. Parental care is typical of these cichlids, although *Cichlasoma urophthalmus* can be said to be especially aggressive towards intruders near the spawning site.

C.urophthalmus is undoubtedly confused with *C.festae*, a South American species that shares the same red basic colour and caudal base spot, although this is larger in *C.urophthalmus*. *Cichlasoma festae* attains a greater body length and is more robust in overall appearance.

Herotilapia multispinosa
Rainbow cichlid
● **Habitat:** Lakes, streams and rivers in Nicaragua and Costa Rica.
● **Length:** 125mm (5in).
● **Diet:** Young specimens will pick at any food which falls to the substrate and show a preference for larval foods, such as bloodworm and gnat larvae, which would be a natural food source in the wild.
● **Sex differences:** Male Rainbow Cichlids are more brightly coloured than females, with

Above: **Cichlasoma urophthalmus**
A male Mexican Tail-spot Cichlid. This species is often confused with C. festae *from South America.*

areas of yellow, brown and red overlaid with a line of black blotches that begins just behind the eye and extends along the mid lateral line. This is punctuated by a large spot or blotch just beyond the middle of the body and finishes with a blotch at the base of the tail. The female shares the same basic pattern, although drab by comparison, and is shorter overall with shorter fins.

● **Aquarium compatibility:** This is perhaps the most compatible of the small Central American cichlids available to aquarists. It is an excellent small dither fish among large *Cichlasoma* species, which will not feel threatened by its diminutive stature in the aquarium.

● **Aquarium breeding:** This species is easy to breed and will spawn when only halfway to adult size! Breeding females change from their usual drab pattern to take on a yellow hue, which appears to be for brood recognition. A pair will accept almost any site in the aquarium for spawning, including bogwood, rocks or plant pots.

Above: **Herotilapia multispinosa**
The Rainbow Cichlid is an ideal species to keep in a small community system. Very easy to breed.

Aquarium-bred specimens are so far removed from the original wild form that the colour pattern is reduced to a brown body with lateral banding or spots.

Neetroplus nematopus
Pygmy Green-eyed Cichlid
● **Habitat:** Lakes and rivers in Nicaragua and Costa Rica. River populations live in fast-flowing waters above rock-strewn substrates.
● **Length:** 75-100mm (3-4in).
● **Diet:** Lake populations are known to be algae grazers and should be offered fine leaf spinach, lettuce or peas. In addition, *Neetroplus* will accept any larval or shrimp food and will thrive even if fed solely on flake food. However, a varied diet soon brings sexually mature fishes into breeding condition.
● **Sex differences:** Adult males appear to develop a slight nuchal hump and usually can be identified by their dominant behaviour. Some aquarists suggest that males exhibit slight finnage extensions in comparison with females and this can be a useful guide.

Below: **Neetroplus nematopus**
One of the smallest yet one of the most aggressive species available.

- **Aquarium compatibility:** All aquarium rules are broken by this tiny Central American cichlid. A 75-100mm (3-4in) male can cause havoc in small and large aquariums alike, and it will not show any fear of large fishes, especially if paired and brood protecting. Although ideally suited to small aquariums, these demons would not come unstuck in a busy large cichlid community!
- **Aquarium breeding:** These are cave- or hole-spawning cichlids. In the aquarium, a breeding pair will accept a small plant pot buried in the gravel (leaving an opening only large enough for them to squeeze through) or a narrow space in a rockwork cluster. Once the eggs have been produced, the female reverses her colour pattern. The body darkens from grey to black, and the characteristic black bar becomes white. (This colour change is one of the most extreme to occur among cichlids.) The male also reverses his colour pattern when brood protection begins. Although up to 100 eggs can be produced in one spawning, about 30-40 is the usual number. Brood-protecting parents will not hesitate to attack intruders, but other community fishes tend to recognize the demon pair as potential trouble to be avoided.

Above: **Neetroplus nematopus**
A male displaying the fin extensions and slight nuchal hump characteristic of the species. These fishes are said to pick parasites off Cichlasoma nicaraguense *in the wild.*

Neetroplus nematopus is noted as a 'cleaner' fish to larger cichlids, much in the same way as the Cleaner Wrasse(Labroides dimidiatus) attends to its fellow marine fishes.

Petenia splendida
Bay Snook; Giant Cichlid
- **Habitat:** Mexico, Guatemala, Belize and Nicaragua. It lives in still waters, such as quiet stretches of large rivers, typically in densely planted areas near waterlogged brush or tree trunks.
- **Length:** 500mm (20in).
- **Diet:** These are avid fish eaters, although this appetite for eating fellow aquarium occupants can be tempered by offering large earthworms, whole shrimp or prawns, maggots and fish pieces as a diet.
- **Sex differences:** Males tend to display longer fin rays and are invariably brighter in colour than the females.
- **Aquarium compatibility:** A full-sized specimen would hardly suit a small community aquarium, although youngsters can be brought up easily among smaller Central American cichlids. In the semi-adult stage, it is ideally suited to a large aquarium system and can be kept with adult Cichlasoma and Cichla species.
- **Aquarium breeding:** Aquarium spawnings have been reported, although this is a rare occurrence compared with the many Cichlasoma species that are regularly spawned. As one of the giant – or true Guapote – species, Petenia splendida is capable of producing up to 5000 eggs in a spawning, although less than half of these will be fertilized. Large rocks or large plant pots (split down the middle and positioned vertically) should be placed in the breeding tank as potential spawning sites.

Juvenile Giant Cichlids have been available in recent years, although never in any regular quantity. Several colour forms exist, including red-eyed brown, black speckled, silver-black speckled, and red.

Above and below: **Petenia splendida** *The Bay Snook in adult colour (above) is a most attractive giant species. Juveniles (below) are more likely to be encountered by fishkeepers, but they are by no means common.*

Index to fishes

Page numbers in **bold** indicate major references, including accompanying photographs. Page numbers in *italics* indicate captions to other illustrations. Less important text entries are shown in normal type.

A

African Rift Valley Lake Cichlid 13
Astronotus ocellatus 71

B

Bandit Cichlid **66-7**
Bay Snook **74-5**
Black Belt Cichlid **53-4**
Black-blotch Cichlid **56-7**
Blue-eyed Cichlid **66-7**
Blue Red-top Cichlid **42-4**
Blue Speckled Cichlid **71**
Blue Texas Cichlid **44-5**, 46

C

Cichla ocellaris 12, **42-3**
Cichlasoma aureum **42-4**
  bifasciatum 30
  biocellatum 61
  carpinte **44-5**, 46
  citrinellum **45-6**, 51
  cyanoguttatum 36, 44, 45, **46**
  dovii 12-3, *12-3*, **46-7**
  ellioti Endpapers
  fenestratum 49
  festae 72
  friedrichsthalii **48-9**, 57
  hartwegi 38, **49**
  intermedium 15, **50**
  labiatum Copyright page, 46, **50-1**
  longimanus **52-3**
  maculicauda 48, **53-4**, 56, 58
  managuense 12, 28, *29*, **54-5**, 57
  meeki 11, *11*, 43, **55-6**, 63
  melanurum **56-7**
  motaguense 12, 28, 48, 49, 55, **56-7**
  nicaraguense 26, 54, **58-9**, 69
  nigrofasciatum 37, **59-61**, 67, 68
  octofasciatum **60-1**
  panamense **62-3**
  robertsoni **62-3**
  sajica **64-5**
  salvini 11, 26, **65-6**
  sieboldii 30, **66-7**
  spilurum 24, *24*, 64, 65, **66-7**, 68
  spinosissimum 64, 65, **68**
  synspilum *26*, 48, 54, 56, 58, **69**
  tetracanthus 46
  trimaculatum **70-1**
  umbriferum **71**
  urophthalmus **71-2**
Cleaner Wrasse 74
Convict Cichlid *37*, **59-61**, 67, 68
Cuban Cichlid 45

D

Dow Cichlid **46-7**
Dow's Cichlid **46-7**

E

Eye-spot Cichlid **42-3**

F

False Blue-eyed Cichlid **68**
False Jade-eyed Cichlid **68**
Firehead Cichlid **69**
Firemouth Cichlid 11, 43, **55-6**, 61, 62
Friedrichsthal's Cichlid **48-9**

G

Giant Cichlid **74-5**
Gold Cichlid **42-4**
Green Texas Cichlid **44-5**

H

Herotilapia multispinosa **72-3**
Hypostomus plecostomus 33

I

Intermedium Cichlid **50**

J

Jack Dempsey Cichlid **60-1**
Jade-eyed Cichlid **66-7**
Jaguar Cichlid 12, *36*, **54-5**

L

Labroides dimidiatus 74
Large-lipped Cichlid **50-1**
Lemon Cichlid **45-6**
Loricariidae 33

M

Managua Cichlid **54-5**
Metallic Green Cichlid **62-3**
Mexican Tail-spot Cichlid **71-2**
Midas Cichlid **45-6**
Motagua Cichlid 55, **56-7**

N

Neetroplus nematopus *10*, 11, 24, 30, 33, 47, 62, 63, **73-4**
Nicaragua Cichlid **58-9**

O

Oscar 36, 71

P

Panama Cichlid **62-3**
Petenia splendida 12, 48, **74-5**
Pimelodella 33
Pterygoplichthys sp. **33**
Pygmy Green-eyed Cichlid **73-4**

Q

Quetzal 69

R

Rainbow Cichlid **72-3**
Red Devil **45-6**, **50-1**
Rhamdia sp. 33
Rose-breasted Cichlid **52-3**

S
Sajie's Cichlid **64-5**
Salvin's Cichlid 61, **65-6**
Shoulder-spot Cichlid **70-1**
Spathiphyllum wallisii 23
Sword Plant 24, 29

T
Tail Bar Cichlid **49**
Texas Cichlid 45

Three-spot Cichlid **70-1**
Tricolor Cichlid 11, **65-6**
True Texas Cichlid **46**

W
Wolf Cichlid **46-7**

Z
Zebra Cichlid **59-61**

Picture credits

Artists
Copyright of the artwork illustrations on the pages following the artists' names is the property of Salamander Books Ltd.

Clifford and Wendy Meadway: 17, 18, 19, 39

Colin Newman (Linden Artists): 24-5, 26-7, 28-9

Photographs
The publishers wish to thank the following photographers and agencies who have supplied photographs for this book. The photographs have been credited by page number and position on the page: (B) Bottom, (T) Top, (C) Centre, (BL) Bottom left etc.

Eric Crichton © Salamander Books Ltd: 16, 32

Arend van den Nieuwenhuizen: 44, 60-1(B), 61(C)

Laurence Perkins: 45

Mike Sandford: 56, 73(B)

David Sands: Copyright page, 10, 15(B), 23, 26, 31, 33, 36(B), 37, 43(T), 46-7(T), 50, 51(T,B), 55(T), 57(B), 62, 67(T), 73(T)

Ian Sellick: 40-1

Uwe Werner: Endpapers, Title page, 11, 12-3, 14-5(T), 22, 25, 29, 30, 36(T), 38, 42, 43(B), 47, 48(B), 48-9(T), 52-3, 54, 57(T), 58-9, 61(T), 63(T,B), 65, 66-7(B), 68(T,B), 69, 70(T,B), 71, 72, 74, 75(T,B)

Rudolf Zukal: 64

Acknowledgements
The publishers wish to thank Philip Murray of Driftwood Aquarium, Chorley, for the loan of several cichlids photographed and illustrated in this book.

Companion volumes of interest:

A Fishkeeper's Guide to THE TROPICAL AQUARIUM
A Fishkeeper's Guide to COMMUNITY FISHES
A Fishkeeper's Guide to COLDWATER FISHES
A Fishkeeper's Guide to MARINE FISHES
A Fishkeeper's Guide to MAINTAINING A HEALTHY AQUARIUM
A Fishkeeper's Guide to GARDEN PONDS
A Petkeeper's Guide to REPTILES AND AMPHIBIANS

Cichlasoma ellioti

PRINTED IN BELGIUM BY
proost
INTERNATIONAL BOOK PRODUCTION